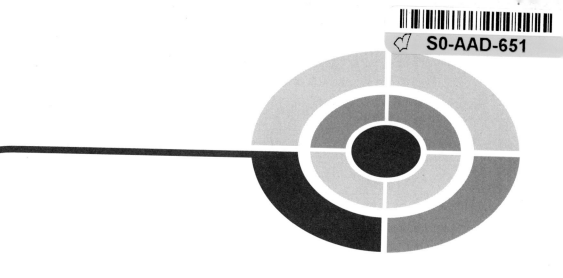

Earth Science Demystified

Demystified Series

Advanced Statistics Demystified
Algebra Demystified
Anatomy Demystified
Astronomy Demystified
Biology Demystified
Business Statistics Demystified
Calculus Demystified
Chemistry Demystified
College Algebra Demystified
Earth Science Demystified
Everyday Math Demystified
Geometry Demystified
Physics Demystified
Physiology Demystified
Pre-Algebra Demystified
Project Management Demystified
Statistics Demystified
Trigonometry Demystified

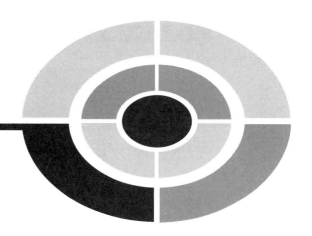

Earth Science Demystified

LINDA WILLIAMS

McGRAW-HILL

New York   Chicago   San Francisco   Lisbon   London
Madrid   Mexico City   Milan   New Delhi   San Juan
Seoul   Singapore   Sydney   Toronto

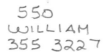

Cataloging-in-Publication Data is on file with the Library of Congress

This book is dedicated to my father, Douglas Williams, who loved nature and was an even bigger "rock hound" than I am. He spent his life pointing out the beauty and wonders of the natural world and honored God through reverence and care of His work. Thank you, Dad. I love you and will never forget.

The sponsoring editor for this book was Judy Bass and the production supervisor was Pamela A. Pelton. It was set in Times Roman by Keyword Publishing Services Ltd. The art director for the cover was Margaret Webster-Shapiro; the cover designer was Handel Low.

Printed and bound by RR Donnelley.

 This book is printed on recycled, acid-free paper containing a minimum of 50% recycled, de-inked fiber.

McGraw-Hill books are available at special quantity discounts to use as premiums and sales promotions, or for use in corporate training programs. For more information, please write to the Director of Special Sales, McGraw-Hill Professional, Two Penn Plaza, New York, NY 10121-2298. Or contact your local bookstore.

CONTENTS

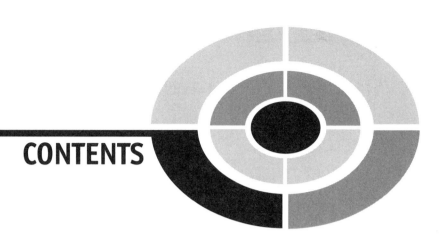

PREFACE

Earth Science is made up of many different areas of geological study. Since the Earth contains everything from clouds (meteorology) and oceans (marine biology) to fossils (paleontology) and earthquakes (geology/plate tectonics), there is a lot to choose from!

This book is for anyone with an interest in Earth Science who wants to learn more outside of a formal classroom setting. It can also be used by home-schooled students, tutored students, and those people wanting to change careers. The material is presented in an easy-to-follow way and can be best understood when read from beginning to end. However, if you just want to brush up on specific topics like minerals and gems or volcanoes, then those chapters can be reviewed individually as well.

You will notice through the course of this book that I have mentioned many milestone theories and accomplishments of geophysicists, oceanographers, seismologists, and ecologists to name a few. I have highlighted these knowledge leaps to give you an idea of how the questions and bright ideas of curious people have advanced humankind.

Science is all about curiosity and the desire to find out how something happens. Nobel Prize winners were once students who daydreamed about new ways of doing things. They knew answers had to be there and they were stubborn enough to dig for them. The Nobel Prize for Science (actors have Oscar and scientists have Nobel) has been awarded over 470 times since 1901.

In 1863, Alfred Nobel experienced a tragic loss in an experiment with nitroglycerine that destroyed two wings of the family mansion and killed his younger brother and four others. Nobel had discovered the most powerful weapon of that time, dynamite.

By the end of his life, Nobel had 355 patents for various inventions. After his death in 1896, Nobel's will described the establishment of a foundation to create five prizes of equal value "for those who, in the previous year, have

contributed best toward the benefits for humankind," in the areas of Earth Science, Physics, Physiology/Medicine, Literature and Peace. Nobel wanted to recognize the heroes of science and encourage others in their quest for knowledge.

Earth Science also has individual prizes and awards specific to geology. The Penrose Medal (pure geology), Crawford Prize (nonlinear science, e.g., dynamics and computations/simulations), and the Day Medal (geophysics and geochemistry) are all awarded in recognition of outstanding Earth Science research and advancements.

My hope is that in learning of the many simple ideas and observations that changed our understanding of the way the Earth functions, you too will be encouraged to let your own creative thoughts tackle ongoing Earth Science challenges.

This book provides a general overview of Earth Science with sections on all the main areas you'll find in an Earth Science classroom or individual study of the subject. The basics are covered to familiarize you with the terms and concepts most common in the experimental sciences like Earth Science. Additionally, I have listed helpful Internet sites with up-to-date and interactive geological information and simulations.

Throughout the text, I have supplied lots of everyday examples and illustrations of natural events to help you visualize what is happening beneath, on, or above the Earth's surface. There are also quiz, test, and exam questions throughout. All the questions are multiple choice and a lot like those used in standardized tests. There is a short quiz at the end of each chapter. These quizzes are "open book." You shouldn't have any trouble with them. You can look back at the chapter text to refresh your memory or check the details of a natural process. Write your answers down and have a friend or parent check your score with the answers in the back of the book. You may want to linger in a chapter until you have a good handle on the material and get most of the answers right before moving on.

This book is divided into major sections. A multiple-choice test follows each of these sections. When you have completed a section, go ahead and take the section test. Take the tests "closed book" when you are confident about your skills on the individual quizzes. Try not to look back at the text material when you are taking them. The questions are no more difficult than the quizzes, but serve as a more complete review. I have thrown in lots of wacky answers to keep you awake and make the tests fun. A good score is three-quarters of the answers right. Remember, all answers are in the back of the book.

The final exam at the end of the course is made up of easier questions than those of the quizzes and section tests. Take the exam when you have finished all the chapter quizzes and section tests and feel comfortable with the material as a whole. A good score on the exam is at least 75% of correct answers.

With all the quizzes, section tests, and the final exam, you may want to have your friend or parent give you your score without telling you which of the questions you missed. Then you will not be tempted to memorize the answers to the missed questions, but instead go back and see if you missed the point of the idea. When your scores are where you'd like them to be, go back and check the individual questions to confirm your strengths and any areas that need more study.

Try going through one chapter a week. An hour a day or so will allow you to take in the information slowly. Don't rush. Earth Science is not difficult, but does take some thought. Just slug through at a steady rate. If you are really interested in earthquakes, spend more time on Chapter 12. If you want to learn the latest about the weather forecasting, allow more time on Chapter 15. At a steady pace, you will complete the course in a few months. After you have completed the course and become a geologist-in-training, this book can serve as a ready reference guide with its comprehensive index, appendices, and many examples of rock types, cloud structures, and global geochemical systems.

Suggestions for future editions are welcome.

Linda Williams

Acknowledgments

Illustrations in this book were generated with CorelDRAW and Microsoft PowerPoint and Microsoft Visio courtesy of the Corel and Microsoft Corporations, respectively.

United States Geological Survey information and maps were used where indicated.

A very special thanks to Dr. Richard Gordon (Plate Tectonics), Sandy Schrank and Abbie Beck (Fossils) for help in editing the manuscript of this book.

Many thanks to Judy Bass and Scott Grillo at McGraw-Hill for your confidence and assistance.

Thank you also to Rice University's Weiss School of Natural Sciences staff and faculty for your friendship, support, and flexibility in the completion of this work.

Many thanks to the folks at Kenny J's and Starbucks, who graciously allowed me to be their resident writer.

My heartfelt thanks to my children, Evan, Bryn, Paul, and Elisabeth for your love and faith. Also, thanks Mom for your constant encouragement and love.

About the Author

Linda Williams is a nonfiction writer with specialties in science, medicine, and space. She has worked as a lead scientist and technical writer for NASA and McDonnell Douglas Space Systems, and served as a science speaker for the Medical Sciences Division at NASA–Johnson Space Center. Currently, Ms. Williams works in the Weiss School of Natural Sciences at Rice University, Houston, Texas. She is the author of the popular *Chemistry Demystified*, another volume in this series.

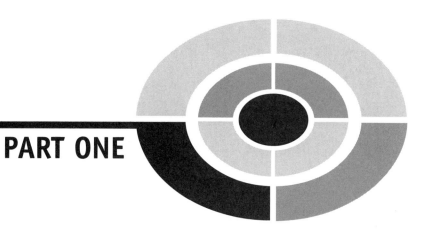

PART ONE

Earth

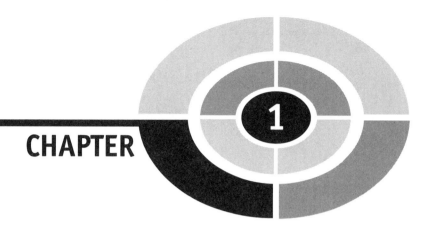

CHAPTER 1

Planet Earth

From space, our world looks like a brilliant blue marble. Sometimes called the "blue planet," the Earth is over 70% water and is unique in our solar system. Clouds, fires, hurricanes, tornadoes, and other natural characters may change the Earth's face at times, but in our solar system, this world is the only one capable of life as we know it.

Native peoples, completely dependent on Mother Earth for everything in their lives, worshipped the Earth as a nurturing goddess that provided for all their needs. From the soil, came plants and growing things that provided clothing and food. From the rivers and seas, came fish and shellfish for food, trade articles, and tools. From the air, came rain, snow, and wind to grow crops and alter the seasons. The Earth was never stagnant or dull, but always provided for those in her care. Ancient people thought Mother Earth worked together with Father Sun to provide for those who honored her.

Today, astronauts orbit the Earth in spaceships and scientific laboratories, 465 km above the Earth, marvel at the Earth's beauty, and work toward her care. Former astronaut Alan Bean communicates this beauty by painting from experience and imagination. Astronaut Tom Jones publishes books for young and old of his space experiences. Other NASA astronauts, scientists, engineers, and test pilots have communicated their wonder and appreciation for our fragile world through environmental efforts that address earth science issues. The study of *geology* includes many areas of global concern.

> **Geology** is the study of the Earth, its origin, development, composition, structure, and history.

But how did it all start? What of the Earth's earliest beginnings? Many scientists believe the Sun was formed from an enormous rotating cloud of dust and gases pulled by gravity toward an ever denser center of mass. The constant rotation flattened things out and allowed debris (some the size of oranges and others the size of North America) to form planets, the Moon, and comets.

The larger pieces of matter in this debris field had enough gravity to grab up small cosmic chunks, glob them together, and allow them to grow larger. When the gathering debris got to be over 350 km across, it was slowly shaped into a sphere by gravity. Figure 1-1 illustrates the steps this formation might have taken.

Other scientists think that everything came about in one gigantic explosion, the Big Bang. Everything was pretty much developed and just simply spiraled out to take the places that we know today. In fact, some astronomers believe that the Universe is expanding. They think all the galaxies are getting further and further apart to almost unimaginable distances. Seems like it would be tough to study something that is moving further away from you all the time!

For the study of Earth Science, though, that is not a problem. The entire planet is a laboratory and provides a lot of great samples.

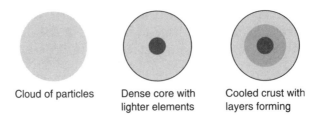

Cloud of particles   Dense core with   Cooled crust with
                       lighter elements   layers forming

Fig. 1-1. Gravity-shaped space debris into a sphere depending on weight and size.

Size and Shape

The shape of the Earth was guessed at for thousands of years. Most early people thought the land and seas were flat. They were afraid that if they traveled too far in one direction, they would fall off the edge. Explorers who sailed to the limits of known navigation were thought to be crazy and surely on the path to destruction. Since many early ships didn't return from long voyages (probably sunk by storms), people thought they had either gone too far and simply fallen off, or had encountered terrible sea monsters and were destroyed.

It wasn't until the respected Greek philosopher, *Aristotle* (384–322 BC), noticed that the shadow cast by the Earth onto the Moon was curved, that people began to wonder about the flat Earth idea. Remember, Aristotle was widely respected in Greece and had written about many subjects including, logic, physics, meteorology, zoology, theology, and economics, so some people wondered if he might be right about the round Earth too. Aristotle believed the Earth was the center of the solar system.

In the early 1500s, Polish astronomer, Nicholas Copernicus, sometimes called the Father of Modern Astronomy, suggested that the Earth rotated around the Sun. His calculations and experiments all pointed to this fact. Unfortunately, many people believed that the Earth was the center of the Universe. They didn't like the idea of the Earth being just another rock circling the Sun. It threatened everything they believed in, from the way they raised crops, to their faith in God. Copernicus and others to follow him, however, continued to question and write about the way things worked and the Earth's place in the cosmos.

It didn't help early people that the Sun, though very bright, doesn't look all that big in the sky. To someone standing on the Earth and seeing fields, mountains, ocean, or whatever, as far as the eye can see, it was no wonder most people thought the Earth was the center of everything. They had no idea of the distance.

The Earth is known as one of the *inner planets* in our solar system. The four *terrestrial* or Earth-like planets found closest to the Sun are Mercury, Venus, Earth, and Mars. They formed closest to the Sun with higher heat than the farther flung planets. Most of the radiation and other solar gases expelled by the Sun blew off high levels of hydrogen, helium, and other light gases to leave behind rock and heavy metal cores. These "hard" planets, including our Moon, are similar chemically and the best picks for establishing human colonies in the near future.

The *outer planets*, made up of volatile matter slung way out into space, are huge compared to the inner planets. These include Jupiter, Saturn, Uranus,

Neptune, and Pluto (the tiny "oddball" of the outer planets made mostly of ice). The giant outer planets have rocky cores, but are mostly made of nebular gases from the original formation of the Sun.

Just as the planets are held in different orbits by the Sun's gravity, the well-defined rings of Saturn made up of gases and particles are also held in orbit by gravity.

To remember the placement of the nine planets in our solar system, picture a baseball field. The distances are nowhere near proportional, but if you think of the inner planets (Mercury, Venus, Earth, and Mars) as the "infield" and the outer planets (Jupiter, Saturn, Uranus, Neptune, and Pluto) as the "outfield," it's easy to keep them straight. Figure 1-2 shows the Earth's place

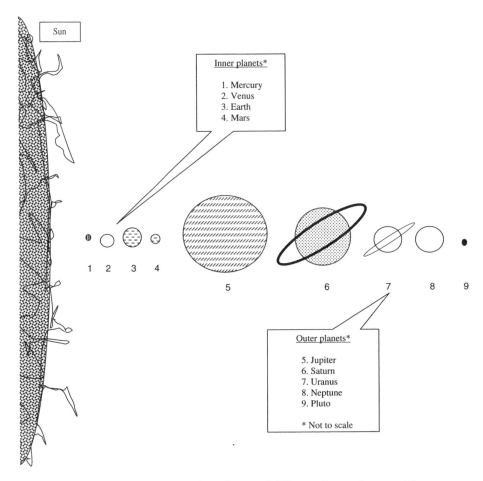

Fig. 1-2. The solar system has planets of different sizes and composition.

in our solar system and gives a rough idea of the different sizes of the planets and the Moon.

Compared to the gigantic Sun, which is over 332,000 times the mass of the Earth, the Earth is tiny, a bit like the size of a human as compared to the size of an ant. The Sun is 1,391,000 km in diameter compared to the Earth which is approximately 12,756 km in diameter. That means the diameter of the Sun is over 100 times the diameter of the Earth. To picture the size difference, imagine that the Sun is the size of a basketball. In comparison, the Earth would be about the size of this "o."

Our planet turns on its axis once a day at a tilt of 23.5° to the plane of the Earth's orbit around the Sun. The other planets spin on their axes as well and roughly share the same plane of rotation as the Earth. The colossal size of the rotating Sun holds the planets in their particular places by gravity.

> The **plane of the ecliptic** is the angle of incline with which the Earth rotates on its axis around the Sun.

The distance to the Sun is an average of 93 million miles from the Earth. This distance is so huge that it is hard to imagine. It has been said that if you could fly to the Sun in a jet going 966 km/hr, it would take over 300 years to get there and back.

Earth's Place in the Galaxy

Even though our Sun seems to be the center of our Universe, it is really just one of the kids on the block. Our solar system is found on one of the spiral arms, *Orion*, of the spiral galaxy known as the *Milky Way*.

> The **Milky Way** is one of millions of galaxies in the Universe. The *Andromeda* galaxy is the nearest major galaxy to the Milky Way.

Think of the Milky Way galaxy as one "continent" among billions of other continents in a world called the Universe. Its spiraling arms or "countries" are called *Centaurus*, *Sagittarius*, *Orion*, *Perseus*, and *Cygnus*. The Milky Way galaxy is around 100,000 light years across. The center of the Milky Way is made up of a dense molecular cloud that rotates slowly clockwise throwing off solar systems and cosmic debris. It contains roughly 200 billion (2×10^{12}) stars.

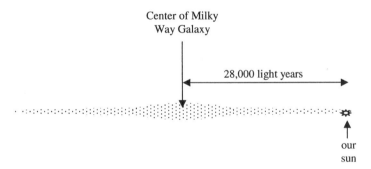

Fig. 1-3.   The solar system is at the edge of the Milky Way galaxy.

Although *Andromeda* is the closest full-size galaxy to the Milky Way, the *Sagittarius Dwarf*, discovered in 1994, is the closest Galaxy. It is 80,000 light years away or nearly 24 kiloparsec. (A parsec is 3.26 light-years away.)

> A **light-year** is a unit of distance, which measures the distance that light travels in one year.

Light moves at a velocity of about 300,000 km/sec. So in one year, it can travel about 10 trillion km. More precisely, one light-year is equal to 9,500,000,000,000 km.

Orion, our "country" within the Milky Way, has many different star systems or "cities." Each star solar system is like a "city" with buildings, parks, and homes. Our solar system is located on the outer edge of the Orion arm. The planets of the solar systems are the "buildings and homes."

Figure 1-3 shows an edge view of the local Milky Way galaxy and our place in it.

Earth's Formation

In 1755, Immanuel Kant offered the idea that the solar system was formed from a rotating cloud of gas and thin dust. In the years since then this idea became known as the *nebular hypothesis*. The clouds that Kant described could be seen by powerful telescopes. The Hubble Space telescope has sent back images of many of these beautiful formations called *nebulae*.

NASA has many images of nebulae photographed from the Hubble Space Telescope. The following websites will give you an idea of the different nebulae that scientists are currently studying:

http://hubble.nasa.gov
http://science.msfc.nasa.gov
www.nasa.gov/home/index.html
http://hubblesite.org/newscenter

The most outstanding of these might be the Horseshoe and Orion nebulae. These beautiful cosmic dust clusters allow space scientists to study the differences between cosmic cloud shapes, effect of gravitational pull, and other forces that influence the rotation of these dust clouds.

It's likely that when the Earth was first forming in our young solar neighborhood, it was a molten mass of rock and metals simmering at about 2000°C. The main cloud ingredients included hydrogen, helium, carbon, nitrogen, oxygen, silicon, iron, nickel, phosphorus, sulfur, and others. As the sphere (Earth) cooled, the heavier metals like iron and nickel sunk deeper into the molten core, while the lighter elements like silicon rose to the surface, cooled a bit, and began to form a thin crust. Figure 1-4 shows the way the elements shaped into a multilayer crust. This crust floated on a sea of molten

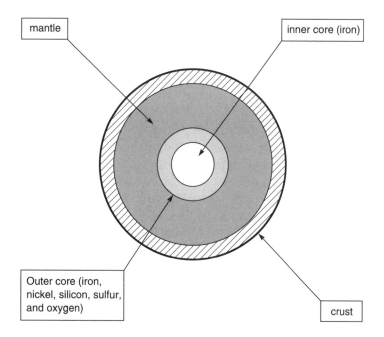

Fig. 1-4. The Earth has four main layers.

rock for about four billion years, sputtering volcanic gases and steam from the impact of visitors like ice comets. Time passed like this with an atmosphere gradually being formed. Rain condensed and poured down, cooling the crust into one large chunk and gathering into low spots, and flowing into cracks forming oceans, seas, lakes, rivers, and streams.

Gravity

If the Earth is spinning, then what force keeps us and everything else in place? *Gravity.*

In 1666, English scientist, Sir Isaac Newton (the guy who had an apple fall off a tree and land on his head) said the objects on a spinning Earth must be affected by centrifugal force. He thought the objects on the Earth would fly off unless there was a stronger force holding them on. This line of thinking led Newton to come up with the *Universal Law of Gravitational Attraction.*

Newton described the law in the following mathematical way:

$$F \text{ is proportional to } \frac{M_1 \times M_2}{d^2}$$

where F is the force of gravitational attraction, M_1 and M_2 are the masses of two attracting bodies, and d is the distance between the center of M_1 and the center of M_2. The larger M_1 and M_2 are, and the smaller d is, then the greater the F (force of attraction) will be. So, since the Earth is huge compared to a horse or a human or volleyball, the force of attraction to the Earth is huge. When planets are heavy and close together, they will be attracted to each other!

Newton also realized that since gravity pulls all objects toward the Earth's center (known as a *radial force*), the centrifugal force (the force of the object pulling away as it spins) is greater the farther away the object from the axis of spin. In other words, the centrifugal force is greatest at the equator and less at the poles. The interaction of the two forces causes the Earth to be flatter at the poles and a bit wider at the waistline (equator). This is measured at the Earth's radius as 6357 km at the poles, but bulges at the equator to 6378 km. The Earth is so big though that it still looks like a perfect sphere from space.

Biosphere

All of life on the Earth is contained in the *biosphere*. All the plants and animals of the Earth live in this layer which is measured from the ocean floor to the top of the atmosphere. It includes all living things, large and small, grouped into *species* or separate types. The main compounds that make up the biosphere contain carbon, hydrogen, and oxygen. These elements interact with other Earth systems.

The **biosphere** includes the hydrosphere, crust, and atmosphere. It is located above the deeper layers of the Earth.

Life is found in many hostile environments on this planet, from extremely hot temperatures near volcanic spouts rising from the ocean floor to polar subzero extremely cold temperatures. The Earth's biodiversity is truly amazing. Everything from exotic and fearsome deep-ocean creatures to sightless fish found in underground caverns and lakes are part of the biosphere. There are sulfur-fixing bacteria that thrive in sulfur-rich, boiling geothermal pools, and there are frogs that dry out and remain barely alive in desert soils until infrequent rains bring them back to life. It makes the study of Earth Science fascinating to people of many cultures, geographies, and interests.

However, the large majority of biosphere organisms that grow, reproduce, and die are found in a narrower range. The majority of the Earth's species live in a thin section of the total biosphere. This section is found at temperatures above zero, a good part of the year, and upper ocean depths to which sunlight is able to penetrate.

The vertical section that contains the biosphere is roughly 20,000 m high. The section most populated with living species is only a fraction of that. It includes a section measured from just below the ocean's surface to about 1000 m above it. Most living plants and animals live in this narrow layer of the biosphere. Figure 1-5 gives an idea of the size of the biosphere.

Atmosphere

The *atmosphere* of the Earth is the key to life development on this planet. Other planets in our solar system either have hydrogen, methane, and ammonia atmospheres (Jupiter, Saturn), a carbon dioxide and nitrogen atmosphere (Venus, Mars), or no atmosphere at all (Mercury).

The atmosphere of the Earth, belched out from prehistoric volcanoes, extends nearly 563 kilometers (350 miles out) from the solid surface of the

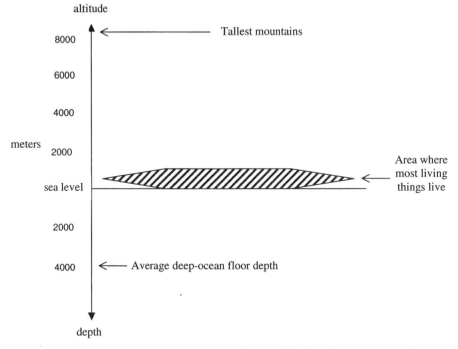

Fig. 1-5. Millions of years ago, the continents of the Earth were in one piece.

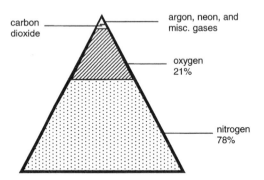

Fig. 1-6. Life exists in a very narrow range.

Earth. It is made up of a mixture of different gases that combine to allow life to exist on the planet. In the lower atmosphere, nitrogen is found in the greatest amounts, 78%, followed by oxygen at 21%. Carbon dioxide, vital to the growth of plants, is present in trace levels of atmospheric gases along with argon and a sprinkling of neon and other minor gases. Figure 1-6 shows the big differences between the amounts of gases present.

Oxygen, critical to human life, developed as microscopic plants and algae began using carbon dioxide in photosynthesis to make food. From that process, oxygen is an important by-product.

The mixture of gases we call, air, penetrates the ground and most openings in the Earth not already filled with water. The atmosphere is the most active of the different "spheres." It presents an ever changing personality all across the world. Just watch the nightly weather report in your own area to see what I mean. In fact, you can see what the weather is doing around the world by visiting the following websites:

www.weather.com
www.theweathernetwork.com
http://www.wunderground.com

We will see all the factors that work together to keep us breathing when we talk about the atmosphere in Chapter 14.

Hydrosphere

The global ocean, the Earth's most noticeable feature from space, makes up the largest single part the planet's total covering. The Pacific Ocean, the largest of Earth's oceans, is so big that all the landmass of all the continents could be fit into it. The combined water of the oceans makes up nearly 97% of the Earth's water. These oceans are much deeper on average than the Earth is high. This large mass of water is part of the *hydrosphere*.

> The **hydrosphere** describes the ever changing total water cycle that is part of the closed environment of the Earth.

The hydrosphere is never still. It includes the evaporation of oceans to the atmosphere, raining back on the land, flowing to streams and rivers, and finally flowing back to the oceans. The hydrosphere also includes the water from underground aquifers, lakes, and streams.

The *cryosphere* is a subset of the hydrosphere. It includes all the Earth's frozen water found in colder latitudes and higher elevations in the form of snow and ice. At the poles, continental ice sheets and glaciers cover vast wilderness areas of barren rock with hardly any plant life. Antarctica makes up a continent two times the size of Australia and contains the world's largest ice sheet.

Lithosphere

The crust and the very top part of the mantle are known as the *lithosphere* (*lithos* is Greek for "stone"). This layer of the crust is rigid and brittle acting as an insulator over the mantle layers below. It is the coolest of all the Earth's layers and thought to float or glide over the layers beneath it. Table 1-1 lists the amounts of different elements in the Earth's crust.

> The **lithosphere** is about 65–100 km thick and covers the entire Earth.

Scientists have determined that around 250 million years ago, all the landmass was in one big chunk or continent. They named the solid land, *Pangea* that means "all earth." The huge surrounding ocean was called *Panthalassa* that means "all seas." But that wasn't the end of the story, things kept changing. About 50 million years later, hot interior magma broke through Pangea and formed two continents, Gondwana (the continents of Africa, South America, India, Australia, New Zealand, and Antarctica) and

Table 1-1   The variety of elements in the Earth's crust make it unique.

Elements of the Earth's crust	%
Oxygen	46.6
Silicon	27.7
Aluminum	8.1
Iron	5
Calcium	3.6
Sodium	2.8
Potassium	2.6
Magnesium	2.1
Miscellaneous	1.4

Laurasia (Eurasia, North America, and Greenland). Scientists are still trying to figure out why the super continents split up, but "hot spots" in the Earth's mantle seem to help things along.

By nearly 65 million years ago, things had broken apart even more to form the continental shapes we know and love today, separated by water.

Crust

The Earth's *crust* is the hard, outermost covering of the Earth. This is the layer exposed to weathering like wind, rain, freezing snow, hurricanes, tornadoes, earthquakes, meteor impacts, volcano eruptions, and everything in between. It has all the wrinkles, scars, colorations, and shapes that make it interesting. Just as people are different, with their own ideas and histories depending on their experiences, so the Earth has different personalities. Lush and green in the tropics to dry and inhospitable in the deep Sahara to fields of frozen ice pack in the Arctic, the Earth's crust has many faces.

CONTINENTAL CRUST

The landmass of the crust is thin compared to the rest of the Earth's layers. It makes up only about 1% of the Earth's total mass. The continental crust can be as much as 70 km thick. The land crust with mountain ranges and high peaks is thicker in places than the crust found under the oceans and seas, but the ocean's crust, about 7 km thick, is denser.

The continents are the chunks of land that are above the level of ocean basins, the deepest levels of land within the crust. Continents are broken up into six major landmasses: Africa, Antarctica, Australia, Eurasia, North America, and South America. This hard continental crust forms about 29% of the Earth's surface and 3% of the Earth's total volume.

Besides dry land, continents include submerged *continental shelves* that extend into the ocean, like the crust framing the edge of a pie. The continental shelf provides a base for the deposit of sand, mud, clay, shells, and minerals washed down from the landmass.

> A **continental shelf** is the thinner, extended edges of a continental landmass that are found below sea level.

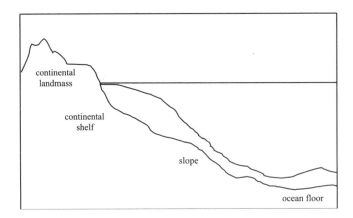

Fig. 1-7. A continental shelf extends the landmass before sloping to the ocean floor.

The continental shelf can extend beyond the shoreline from 10 to 220 miles (16–320 km) depending on location. The water above a continental shelf is fairly shallow, between 200 and 600 feet deep (60–180 km), compared to the greater depths at the slope and below. There is a drop off, called the continental slope, that slips away suddenly to the ocean floor. Here, the water reaches depths of up to 3 miles (5 km) to reach the average level of the seafloor. Figure 1-7 shows the steady thinning of the continental landmass to the different depths of the ocean floor.

A "land" or "dry" continent has more variety than its undersea brother, the oceanic crust, because of weathering and environmental conditions. The continental crust is thicker, especially under mountains, but less dense than the "wet crust" found under the oceans. Commonly, the continental crust is around 30 km thick, but can be up to 50–80 km thick from the top of a mountain.

The continental crust is made up of three main types of rock. These are: *sedimentary*, *igneous*, and *metamorphic* rock. We will learn more about these rock types in later chapters.

OCEANIC CRUST

The land below the levels of the seas is known as the *oceanic crust*. This "wet" crust is much thicker than the continental crust. The average elevation of the continents above sea level is 840 m. The average depth of the oceans is about 3800 m or $4\frac{1}{2}$ times greater. The oceanic crust is roughly 7–10 km thick.

Though not changed by wind and rain as is the continental crust, the oceanic crust is far from dull. It experiences the effect of the intense heat and pressures of the mantle more than the continental crust, because the oceanic crust covers more area.

Even slow processes like sediment collection can trigger important geological events. This happens when the build up of heavy sediments onto a continental shelf by ocean currents causes pieces to crack off and slide toward the ocean floor like an avalanche. When this takes place, the speed of the shift can be between 50 and 80 km/hr. The sudden movement through the water causes intense *turbidity currents* that can slice deep canyons along the ocean floor. We will learn more about ocean currents in Chapter 13.

RIDGES AND TRENCHES

In the middle of the Atlantic Ocean is a north to south mountain range called the Mid-Atlantic Ridge. This ridge is made of many layers of cooled, pushed-up rock from inner crustal depths that have been broken and lifted to form a 16,000 km seam that stretches from Greenland to Antarctica.

Similarly, the East Pacific Ridge contains peaks or seamounts of flattened, dead volcanoes called *guyouts*. These ancient volcanoes were 3660 m above the water level originally, but were eroded down over time by waves crashing against them. Now they are found 1500 m below the waves of the Pacific.

The oceans also contain deep, narrow cuts known as *trenches* that stretch for thousands of miles. Trenches are formed when layers of the crust slam into each other and instead of pushing up like the ridges, they fold at a seam and slide further downward into the layer below. The largest of these trenches, the Mariana, is found in the eastern Pacific.

The Mariana Trench is the deepest trench of this kind on Earth. Located in a north/south line east of the Philippines, it descends over 11,000 m downward and slowly gets deeper. Compared to the height of Mount Everest, the tallest peak on the Earth at 8850 m, the Mariana Trench is gigantic. All of Mount Everest could fit into the Trench with nearly 2200 m of ocean above it to the waves on the surface.

It is $5\frac{1}{2}$ times deeper than the Grand Canyon which is an average of 5000 m deep. We will learn more of this folding action in Chapter 4, when we study plate movement.

It is no wonder the Mariana Trench has been the subject of several science fiction films. It excites the imagination to think about what amazing mysteries of nature might still be discovered at such tremendous depths.

Mantle

The mantle is the next layer down in the Earth's crust. It is located just below the lithosphere. The mantle makes up 70% of the Earth's mass. It is estimated to be about 2900 km thick. The mantle is not the same all the way through. It is divided into two layers, the upper mantle or *asthenosphere* (*asthenes* is Greek for "weak") and the lower mantle. Figure 1-8 shows how the upper and lower mantle layers are separated. These layers are not the same. They contain rock of different density and makeup.

> The highest level of the mantle is called the **asthenosphere** or **upper mantle**. It is located just below the lithosphere.

Except for the zone known as the asthenosphere, the mantle is solid, and its density, increasing with depth, ranges from 3.3 to 6 g/cm^3. The upper mantle is made up of iron and magnesium silicates. The lower part may consist of a mixture of oxides of magnesium, silicon, and iron. This layer is made up of mostly 11 elements: oxygen, silicon, aluminum, iron, calcium, sodium, potassium, magnesium, titanium, hydrogen, and phosphorous. These 11 elements combine with different compounds to form minerals. We will study minerals and gems in depth in Chapter 9.

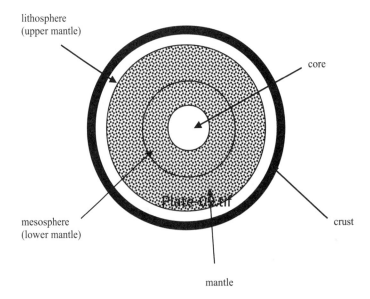

Fig. 1-8. The mantle contains upper and lower layers of different rock types.

The upper mantle is a lot thinner compared to the lower mantle. It can be found between 10 and 300 km below the surface of the Earth and is thought to be formed of two different layers. The bottom layer is tough semi-solid rock and probably consists of silicates of iron and magnesium. The temperature of this layer is 1400–3000°C and the density is between 3.4 and 4.3 g/cm^3. The upper layer of the outer mantle is made up of the same material, but is harder because of its lower temperature.

The upper mantle is solid, but can reach much greater depths than the overlying lithosphere. Compared to the crust, this layer is much hotter, closer to the melting point of rock.

Heat and pressure allows malleability within the mantle. Mantle material moves within this moldable, under layer. Movement is a very slow process, more of a creeping than an actual flowing movement. In Chapters 3, 11, and 12, we will discuss the Earth's layers, volcanoes, and earthquakes in much greater detail which will explain the different ways the Earth's crust shifts and releases stored magma deep within the mantle.

> **Creep** is the extremely slow atom by atom movement and bending of rock under pressure within the mantle.

The heated materials of the asthenosphere become less dense and rise, while cooler material sinks. This works very much like it did when the planet was originally formed. Dense matter sank to form a core, while lighter materials moved eventually upward.

The lower part of the mantle or *mesosphere* is measured from the Earth's core to the bottom of the asthenosphere, at roughly 660 km. Although the average temperature is 3000°C, the rock is solid because of the high pressures. The inner mantle is mostly made up of silicon and magnesium sulfides and oxides. The density is between 4.3 and 5.4 g/cm^3.

> The **mesosphere** is the lower layer of the mantle that borders the Earth's molten core.

The different amounts of heating in the upper and lower parts of the mantle allow solid rock to creep one atom at a time in a certain flow direction. When solids move like this, it is known as *plasticity*. As plasticity occurs in the mantle, slow currents are formed. The continental and oceanic crusts are subducted into the mantle and moved depending on the direction of this deep movement.

Core

Found beneath the mantle is the very center of the Earth. It is made up mostly of iron with a smattering of nickel and other elements. Under extreme pressure, the core makes up about 30% of the total mass of the Earth. It is also divided into two parts, the inner and outer core.

The core is the center part of the Earth and is actually divided into an outer core and inner core. Seismological research has shown that the core has an outer shell of about 2225 km thick with an average density of 10 g/cm^3. The inner core, which has a radius of about 1275 km, is solid with an average density estimated to be 13 g/cm^3. Temperatures in the inner core are estimated to be as high as 6650°C.

The measurement of earthquake waves has suggested that the outer core is fluid and made of iron, while the inner core is solid iron and nickel. The solid center, under extremely high pressure, is unable to flow at all.

MAGNETOSPHERE

The Earth acts as a giant magnet with lines of north/south magnetic force looping from the North Pole to the South Pole. Ancient sailors noticed and used this magnetism to chart and steer a course. Their earliest compasses were just bits of magnetic rock, called *magnetite*, placed on a piece of wood floating in a dish of water. These adventurers knew with tested certainty that every single time the stone was moved to a different direction, its north end would return to point to true north. They didn't know why, but trusted their lives to this knowledge.

The *magnetosphere* is the region of space to which the Earth's magnetic field is limited by the solar wind particles, also called *solar plasma*, blowing outward from the Sun and stretching to distances of over 60,000 km from the Earth. Solar plasma, a gaseous matter made up of freely moving ions and electrons, is electrically neutral overall. It is created in the solar atmosphere (corona) and is continuously blowing outward from the Sun into the solar system. Since the first spacecraft and satellite orbits around the Earth, nearly 50 years ago, a lot has been learned about the interaction between the magnetosphere and the solar wind.

The magnetic field around the Earth is formed by the rotation of the inner core as a solid ball, the different currents in the liquid outer core, and the slower currents of the mantle.

The Earth's magnetic field is kept going by this circulation of molten metals in the core. Scientists believe that the iron–nickel core and its ever moving energy changes into electrical energy. Extreme heat and chemical interactions increase electrical currents and magnetism. The Earth's spin about its axis controls currents and creates the magnetic poles.

Smaller currents, called *eddies*, have an added effect that are thought to bring about the switch in the magnetic rotation. Currently, this magnetic rotation is moving counter-clockwise, but about every million years, something makes it change and rotate in the opposite direction. The polar magnetic current is called the *magnetosphere*. Figure 1-9 shows the powerful circulation of magnetic currents surrounding the Earth.

> The **magnetosphere** extends far beyond the Earth's atmosphere out into space.

The magnetic poles and the geographic north and south poles aren't in the same place. The geographic top and bottom points of the globe are always in the same place, but the magnetic poles move around. Currently, the magnetic pole appears to be moving at a rate of 15 km per year. The magnetic North Pole today, is in the Canadian Arctic between Bathurst and Prince of Wales Islands or about 1300 km from the geographic North Pole.

The South Pole moves around too. It is most recently located off the coast of Wilkes Land, Antarctica roughly 2550 km from the geographical South Pole.

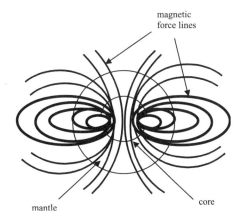

Fig. 1-9.   The magnetosphere of the Earth extends from the north and south poles.

MAGNETOMETERS

The location of the magnetic poles can be figured out from the study of rocks with magnetic particles. The rocks' particles are still aligned with the magnetic poles that existed when they were formed. From studying these rocks, scientists have learned that the magnetic North Pole has moved, over the past 500 million years, from just north of the Philippines in the Pacific Ocean to its more northern location today. Actually the poles are in the same place, but the crust's movement makes their locations appear to migrate like birds.

The magnetic characteristics of underground formations can be measured to figure out geological and geophysical information. This is done through the use of *magnetometers*, which are instruments that measure small differences in the Earth's magnetic field. The first magnetometers were big and bulky, and could only survey a small area. In 1981, however, NASA launched a satellite, equipped with magnetometer technology. This satellite, known as *MAGSAT*, could take magnetic measurements on a continental scale. It allowed geologists to study underground rock formations and the Earth's mantle. MAGSAT also provided clues to landmass movements and the location of deposits of natural gas, crude oil, and other important minerals.

GRAVIMETERS

Geophysicists also measure and record the difference in the Earth's gravitational field to better understand underground structures. Various underground formations and rock types have different effects on the Earth's gravitational field. By measuring minute differences between formations, geophysicists can study underground formations and get a clearer idea of what types of formations lie below ground, and if they contain resources like natural gas.

If sailors were around to navigate the Earth's oceans millions of years ago, their compasses would have still pointed to the magnetic North Pole or "true north." However, they would have sailed to entirely different places on the wandering and shifting crust than they would have today when following the same compass. The Earth is just not the same as it was millions of years ago. A geologist's job is to figure out how it has changed and try to predict what it will do in the future.

Now that you have a general idea of the birth and characteristics of our home planet, let's study the forces that have continued to shape the Earth

since those early days. In Chapter 2, we will learn more about what scientists think might have happened during the early history of our planet when it first became solid and later began to develop character.

Quiz

1. What percentage of water covers the Earth's surface?
    (a) 40%
    (b) 50%
    (c) 70%
    (d) 80%

2. Aristotle was the first person to notice that
    (a) the Moon was round
    (b) mice always live near grain barns
    (c) bubbles appear in fermenting liquids
    (d) the Earth's shadow on the Moon was curved

3. What is the nearest major galaxy to the Milky Way?
    (a) Orion
    (b) Draco
    (c) Andromeda
    (d) Cirrus

4. When space debris measures about 350 km miles across it
    (a) breaks apart
    (b) forms a sphere
    (c) loses its crust
    (d) becomes a meteor

5. The magnetic pole is
    (a) constantly moving
    (b) located exactly at the geographical pole
    (c) only observed in the southern hemisphere
    (d) based on observations of the tides

6. The lithosphere is
    (a) located below the ionosphere
    (b) the crust and very top part of the mantle
    (c) roughly 5–20 km thick
    (d) fluid and soft in all areas

7. When solids flow, it is known as
   (a) flexibility
   (b) plasticity
   (c) a mess
   (d) the magnetosphere

8. The Sun is approximately how many million miles away from the Earth?
   (a) 54
   (b) 75
   (c) 93
   (d) 112

9. The biosphere includes the
   (a) hydrosphere, crust, and atmosphere
   (b) oceans and trenches
   (c) crust, mantle layer, and inner core
   (d) hydrosphere and lithosphere

10. The polar magnetic current is
    (a) very limited in area
    (b) only found around the equator
    (c) the true north of a sailor's compass
    (d) called the magnetosphere

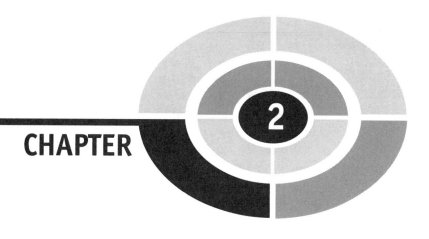

CHAPTER 2

Geological Time

Have you ever thought much about time? Or how long it takes you to do things? How much time it takes to brush your teeth? How long it takes to bake a cake? What is the time difference between riding a bicycle to school instead of walking? How long before your next birthday? How long before your brother finds out you ate the last slice of pizza? What about the amount of time before you get your driver's license or graduate from high school or start college?

These measurements of time are all common within our daily activities, but what about larger amounts of time? How long will it take before you graduate from college and/or graduate school and start a career? How long before you finish a tour of duty in the Armed Services or Peace Corps? How long before you get married, have children, and grandchildren? How long before humans build a colony on the Moon or Mars or beyond? These things could take decades or even a century or two.

What about travel to distant stars? Without a new, as yet undiscovered fuel to travel faster than the speed of light or the "warp engine" of science fiction that travels through bends and folds of time, travel much beyond our solar system is not practical. It can only be done with current rocket engines if the travelers didn't want to return. Generational ships that carried families into space on a grand adventure of colonization would also face radiation shielding, life support/environmental issues, micrometeor impacts, physiological adaptations, and many other challenges.

But does the human race have any other choice? From a scientific view, millions of years from now, the Sun will run out of energy and will eventually cease to exist along with most of the planets in our solar system, including the Earth. But that is a bit far out to plan for, so we might as well keep working on the geological problems we have now. To study and learn about our planet is much more fun!

Earth Time

What about time measured only in our imaginations? What about millions and billions of years? What kind of timescale can bridge vast stretches of time?

> Time that spans millions of years is known as **geological time**. The entire history of the Earth is measured in geological time.

Geological time includes the history of the Earth from the first hints of its formation until today. Geological time is measured mathematically, chemically, and by observation.

Figure 2-1 shows a geological time clock with one second roughly equal to one million years.

In 1785, Scottish scientist James Hutton, called the *father of modern geology*, began to try to figure out the Earth's age from rock layers. He studied and tested local rock layers in an attempt to calculate time with respect to erosion, weathering, and sedimentation.

Hutton knew that over the period of a few years, only a light dusting of sediments are deposited in an undisturbed area. He thought that sedimentary rock that has been compacted and compressed, tighter and tighter, from the weight of upper rock layers must have happened over many ages. He also thought that changes in the sedimentary rock layer, through uplifting and fracturing of weathering and erosion, could only have taken place over a very long period of time. Hutton was one of the first scientists to suggest that the Earth is extremely ancient compared to the few thousand years that earlier theories suggested. He thought that the

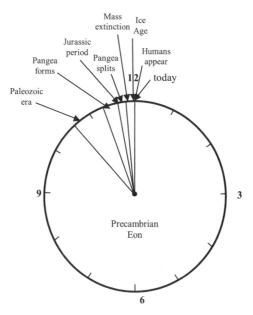

Fig. 2-1. Over 99.9% of the Earth's development happened before humans appeared.

formation of different rock layers, the building of towering mountains, and the widening of the oceans had to have taken place over millions of years.

Hutton wrote the *Principle of Uniformitarianism* that suggested that changes to the Earth's surface happened slowly instead of all at once. His early work paved the way for geologists to consider that the Earth was not in its final form, but was still changing. Gradual shifting and compression changes were possible across different continental land forms.

Time Measurements

Ancient people, until about the 17th century, believed the Earth was approximately 6000 years old. This estimate, based mostly on the history of humankind handed down through stories and written accounts, seemed correct. Except for theory, there were no "scientific" ways to check its accuracy. However, in the 1800s scientists began to test rock samples for their age. It was during this time that scientists used dating methods to suggest the Earth might be millions or even billions of years old.

Relative Time

Geological time is studied by two different methods. The first is a hands-on inspection of the positioning of the different layers of the Earth. This is known as *relative* or *chronostatic time* measurements. Relative time measurements are used to find the age relationships between layers and samples. Using relative time measurements, the age of earth layers is found by comparing them to neighboring layers above and below.

Even when the exact date of rock or materials is not known, it is possible to figure out the sequence of events that led to the current position of a sample. This ordering of samples and events is known as *relative dating*.

> Placing a sample in an approximate time period compared to other samples with known ages is called **relative dating**.

The earliest attempt to order geologic events was done by *Nicolaus Steno* in 1669 when he described the following three laws that placed samples in time:

1. Law of Superposition,
2. Law of Original Horizontality, and
3. Law of Lateral Continuity.

The first law is the simple description of layers as they were piled on top of each other over time. Figure 2-2 shows the simple layering in the *Law of Superposition* that occurs when layers are left undisturbed.

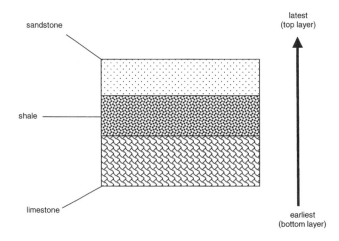

Fig. 2-2.   The oldest rock layers are found below younger layers in the Law of Superposition.

This is the foundation of all geological time measurements. For example, when archeologists study layers of ancient settlements and cities, they record the most recent top layers first, followed by older layers that are uncovered the deeper they dig.

Steno's second idea, the Law of Original Horizontality, was also a new idea at the time. He believed that *sediments* were geologic layers found mostly in a flat, horizontal direction. Figure 2-3 illustrates how uneven layers are still horizontal even after base-layer bending and folding has taken place.

> Any solid material (rock or organic) that settles out from a liquid is known as **sediment**.

Driving along highways that have been cut into hills and mountainsides, you will see horizontal rock layers shifted at steep angles. These sediment layers were shifted after the original sedimentation took place.

The third of Steno's laws describes the *Law of Lateral Continuity*. This law describes the observation that water-layered sediments thin out to nothing when they reach the shore or edge of the area where they were first deposited. This happens even though they were originally layered equally in all directions.

Sometimes scientists find in studying sediments that layers of different sections are missing. These layers have been split far apart through geological movements or by timeless erosion. If a sample is taken from a section with a missing or eroded layer, the true picture of its sedimentation can't be seen.

An *unconformity* is a surface within several layers of sediment where there is a missing sedimentary layer. This is usually found between younger and older rock layers. If this unconformity happens in a wide area of erosion, maybe over an entire mountain range, the time period under study may be misunderstood or completely lost. We will learn more about different kinds

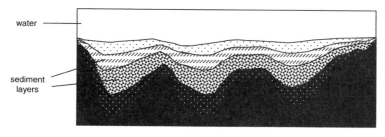

Fig. 2-3. The Law of Original Horizontality describes the overall flat-layered deposition of sediments.

of unconformities found in sedimentary rock when we take a closer look at strata and land eras in Chapter 5.

A *disconformity* contains an area where the sedimentary rock layers, located above and below, are aligned in parallel, but have an area in-between that is different. This can happen when layers of water-covered sediment are uncovered for a time in one era, allowing the environment to erode away or add a different layer, then recovered with water. The layer that was exposed to the environment will be different from its neighbor. However, when they are both covered with water again, the sediment layer will be deposited on both the same way. So unless scientists look closely or the exposed layer is especially thick, the differences might not be noticed.

FOSSILS

Scientists who study ancient plants, animals, and their environment are called *paleontologists* or *paleobiologists*. By studying layers of sedimentary rock and their contents, biologists can tell a lot about what was going on during different geological time periods. The weathering of different rock layers made people think that the Earth must be very old, since it takes a long time for rocks to wear away by water and wind.

A certain climate can be researched by looking for fossilized plants that lived in a specific climate during a certain time period.

Much of the study of relative time measurements is done using *fossil* remains found in rock and frozen in polar ice. Some of these fossils are discovered all in one piece when a road or the foundation of a building is dug, but sometimes only the tip of the toe is showing with the rest buried in rock. Other times, the fossils are like what you see at the beach; a mixture of broken shell and bone fragments scattered across a wide area. Depending on what you are looking for, either of these finds give the paleontologists important information. In fact, when you look at a chart of major fossil types from a certain period, you will only see representative plants or animals. These are the main specimens that have been discovered and placed in a certain time frame. But, the important thing to remember is that just like today, the ancient world had a rich, diverse plant and animal community. You see more than one dog and one pine tree in a large area.

> **Fossils** include the outer cast, imprint, or actual remains of a plant or animal preserved in rock.

Fossils were formed over millions of years by the buildup of sand, silt, or clay over plant and animal specimens in ancient lakes, seas, and oceans. For example, fossils can be the outer shell of a creature like the ancestors of horseshoe crabs called *trilobites*. Fossilized ants and insects, trapped millions of years ago, are often found intact in a petrified, rock-hard tree sap, called *amber*. Depending on the type of tree and environmental conditions at the time, ambers are found in hues from brown and burgundy to orange and pale yellow.

Other fossils are formed beneath frozen tundra, desert sands, and tar pits. Animals that die in a place where the amount of oxygen is very low, like tar pits, suffer very little decay. These are usually found whole, like dinosaur remains.

Fossils are formed and preserved when their soft parts decay and the elements of their hard or bony parts are exchanged for the minerals found in the surrounding sediment layers in which they were first enclosed, like petrified wood or sea life.

Many fossils and seashells are found in areas that were originally primitive seas. Larger fossil remains are most often uncovered by water and wind erosion. Scientists who study and dig at one fossil site one year and then leave it to return a few years later, often find a bit of bone sticking up through the soil that has become exposed by weathering. When paleontologists investigate this new find and uncover the entire skeleton just below the surface, they often begin their work at the site all over again.

This has happened at many of the major fossil sites. A good example is the Burgess Shale quarry in the Canadian Rockies of southern British Columbia, Canada. From the time Charles D. Walcott, working for the Smithsonian Institute, began studying the Cambrian fossils preserved in the shale in the summer of 1909, until the present, many groups from leading museums and universities have worked that site and found new specimens.

One thing to keep in mind is that as the Earth was going through major changes, like the break up of the supercontinent, Pangea, during the Triassic period, the inhabitants of those landmasses were being shifted around as well. During the Mesozoic period, the continents began to drift apart, expanding the oceans between them.

Over millions of years, as the continents moved farther apart, life on the "islands" of continental land got more and more different from their relatives. That is why we see such an unusual diversity of plants and animals in Australia. Without certain predators, living on other continents to sneak up on them and enjoy them as a tasty morsel, animals (like the kangaroo and duck-billed platypus) were able to enjoy the good life and develop in unique ways.

We will take a look at several significant fossil periods and locations in much greater detail in Chapter 10. The huge number and variety of fossils from different time periods and in different environments around the world is amazing. Some areas have produced fossils, large and small, in the tens of thousands.

The world's oceans have also had their share of change.

The sea floor of the Atlantic and Indian oceans appears to have been created at the ocean ridges since the Jurassic period. Geologists have not found any fossils older than the Jurassic period in these sedimentary areas of the sea floor. All the rock of the ocean floors is fairly young in geological time. New rock is created by volcanic eruptions and rising magma.

Ocean levels have changed many times over the years with the highest levels taking place in the Precambrian, Jurassic, and Tertiary periods. During some periods, the oceans were at "high tide" and left deposits of marine organisms far inland. Millions of years ago, a shallow sea stretched from Canada down through the United States. Now that this sea is gone, geologists find exposed rock containing millions of ocean organisms in these areas.

When the oceans receded, the main places of fossil deposit were off the continental land and in the open ocean.

Absolute Time

The second method to study geological time is done by the chemical and radiological testing of different isotopes (forms of the same element) within rock and mineral samples. This is called *absolute* or *chronometric* time measurements.

By using rock and fossil samples that have been classified as to their relationship to one another, laboratory testing can then determine a sample's age and time placement. The two methods work together to give scientists an accurate time picture of a sample's age.

RADIOACTIVITY

The radioactive properties of different elements were discovered in 1896, by Antoine Becquerel when he discovered that a photographic plate in his lab, never exposed to sunlight, had somehow become exposed. The only possible culprit was a nearby uranium salt sitting on the laboratory bench top.

The term, *radioactivity*, was first used by French scientist, Marie Curie, in 1898. Marie Curie and her physicist husband, Pierre, found that radioactive particles were emitted as either electrically negative (−) called *beta* (β) particles, or positive (+) called *alpha* (α) charged particles.

> **Radioactivity** is the characteristic of an element to change into another element through the loss of charged particles from its nuclei.

Following the further understanding and discovery of radioactive breakdown products, researchers began to see a use for radioactivity and radioactive elements, in the study of rock, mineral, and fossil samples.

NUCLEAR REACTIONS

Most chemical reactions are focused on the outer electrons of an element, sharing, swapping, and bumping electrons into and out of the joining elements of a reaction. Nuclear reactions are different. They take place within the nucleus.

There are two types of nuclear reactions. The first is the *radioactive decay* of bonds within the nucleus that emit radiation when broken. The second is the "billiard ball" type of reactions, where the nucleus or a nuclear particle (like a proton) is slammed into by another nucleus or nuclear particle.

RADIOACTIVE DECAY

A radioactive element, like everything else in life, decays or ages. When uranium decays over billions of years, it goes through a process of degrading into lower and lower energy forms until it settles into one that is stable.

The ages of the most ancient rocks can be found by measuring the decay of specific isotopes that are not stable, but break down to other element forms. The sample is dated using testing techniques known as *radiometric dating*. This considers all the various melting and environmental influences that have affected the sample.

When a radioactive element decays, different nuclear particles are given off. These speeding radiation particles can be separated by an electric (magnetic field) and detected in the laboratory:

beta (β) particles = negatively (−) charged particles

alpha (α) particles = positively (+) charged particles

gamma (γ) particles = electromagnetic radiation with no overall

charge (similar to x-rays)

The age of geological samples is found by measuring *radioisotope decay*. Decay of radioactive isotopes is affected by the stability of an element at a certain energy level (where its electrons are stable and bonded). Bismuth (Bi) is the heaviest element in the Periodic Table with a minimum of one stable isotope. All other heavier elements are radioactive.

Geologists study *isotopes* of different chemical elements to find the rate of decay over time. Depending on the rate of decay of a sample, an estimate of its age can be done. It is also possible to find and compare the radioactive decay of a sample to the rock in which it was found. This gives geologists another clue as to the life cycle and history of the specimen, as well as a hint as to how it was deposited.

> **Isotopes** are chemically identical atoms of the same element that have different numbers of neutrons and mass numbers.

A rock sample's age can be found by comparing three pieces of information:

1. the amount of the original element or *parent isotope*,
2. the amount of the new element or *daughter isotope* formed, and
3. the rate of decay of a specific radioactive isotope present in the rock.

A *mass spectrometer* is an instrument that measures the ratios of isotopes in samples. Uranium has all radioactive isotopes while potassium has only one. By noting the rate of decay that uranium-238 displays while losing electrons and alpha particles and trying to become the more stable lead-206, scientists then test other radioactive samples with a similar rate of decay to the stable lead form. Meteorites have been found to be as old as the Earth and older by using this method.

HALF-LIFE

All radioactive isotopes have a specific set, *half-life*. These time periods are not dependent on pressure, temperature, or bonding properties.

> The **half-life** of a radioactive isotope is the time needed for $\frac{1}{2}$ of a specific element sample to decay.

For example, the half-life of $^{238}U_{92}$ is 4.5×10^9 years. It is amazing to think that the uranium found today will be around for another four billion years.

In 1953, Clair Patterson and Friedrich Houtermans separately determined the age of the Earth and the solar system as being around 4.6 billion years old by finding and comparing the radioactive decay rates of isotopes of lead in the earliest rocks known to exist.

Through the radiometric dating of ancient rocks, the Earth was calculated to be over four billion years old. Zircon crystals found in Western Australia have radiometric ages of over 4.3 billion years.

By figuring out how long the oldest lead ores took to change from their earliest formation (nebular gas) to later compression and inclusion in the Earth's crust, scientists were then able to estimate the age of lead-containing meteorites. These meteorites have been dated at nearly 4.6 million years using the radioactive decay of uranium-235 to lead-207.

In this same way, Patterson and Houtermans were able to estimate the age of our solar system to be about 4.54 billion years. The age of our Milky Way galaxy was judged to be between 11 and 13 billion years.

In a study published in *Science* in January 2003, a team of researchers estimated that the Universe was between 11.2 and 20 billion years old.

Most estimates of the Universe's age, in recent years, have ranged between 10 and 15 billion years. Data supplied by the Hubble Space Telescope in 2003 led to a refined estimate of 13–14 billion years.

The new calculations, by Lawrence Krauss of Case Western Reserve University and Brian Chaboyer at Dartmouth College, involved new information about old star clusters in our galaxy and a better understanding of how stars evolve. It was based on when stars are thought to end the main sequence of their lives, a point at which they've used up the hydrogen that fuels thermonuclear fusion and therefore begin to dim.

CARBON DATING

Isotopes are also used in the dating of ancient soils, plants, animals, and the tools of early peoples. An isotope of carbon, ^{14}C, which has a half-life of 5730 years, can be used to calculate geological age. Since the radioactive decay rate of carbon is constant, observing its decay rate allows the measurement of the number of years that have past compared to carbon's half-life.

The preservation of the original organic sample can affect carbon dating. Carbon-14, which decays to ^{14}N, is mostly used for dating samples of fairly recent geological age. Most scientists believe that carbon dating is only accurate for dating specimens thought to be between 30,000 and 50,000 years old. Carbon dating is particularly helpful when finding the age of bone,

Table 2-1   Isotopes used to date geological samples.

Isotopes (original)	Isotopes (decay products)	Half-life (years)	Dating accuracy (range in years)
Carbon-14	Nitrogen-14	5730	30,000–60,000
Rubidium-87	Strontium-87	47 billion	10 million–4.6 billion
Potassium-40	Argon-40	1.3 billion	50,000–4.6 billion
Uranium-238	Lead-206	4.5 billion	10 million–4.6 billion

wood, shell, fossils, and other organic samples since they all contain carbon. These plants and animals use carbon in their basic structure and usually have a good amount of carbon left to be measured.

Other radiometric methods that make use of uranium, lead, potassium, and argon measure much longer time periods since they are not limited to the organic remains of prehistoric samples. Table 2-1 shows a few of the elements used to date samples of different ages.

Dating samples is not an exact science. A lot of factors have to be considered when a sample is dated, including its preservation and the amount of erosion and exposure it has suffered. For example, when lead is depleted from a rock sample through erosion, then the uranium used to date the sample through the breakdown of the uranium to the lead end product (daughter) would show an incorrectly young age.

GEOCHRONOLOGICAL UNITS

If you visit a Museum of Natural Sciences or Natural History you won't be looking at brightly colored paintings or finely crafted statues created by human artists. Instead, you will see thousands of fossils and preserved shells and bones of ancient marine life, plants, and animals created by Nature.

These exhibits provide a chronological history of the Earth by displaying rock, plant, and animal specimens. In addition, besides giving the location where each sample was discovered, most museums also date samples according to their place in geological time.

Geological time is measured and divided into various parts called **geochronological units**.

Although people sometimes use the word eon to mean a really long time, like "it has been eons since I visited with my cousin," the term actually comes from the Greek word, *eos*, meaning "dawn." Geochronological units start with the major divisions of time called *eons*. Eons are measured in millions of years. As time dating is refined, the boundaries of the eons may change slightly, but most geological dating is calculated with an error of margin ± 60 million years either way.

The three major *eon* divisions are the *Archean*, *Proterozoic*, and *Phanerozoic*.

The *Archean* (Greek for "ancient") eon is commonly thought to include the oldest rocks known. It is often called the early Precambrian era which begins with the formation of the Earth, about four billion years ago until about 2500 million years ago. The Proterozoic eon is now thought of as being part of the late Precambrian era.

Figure 2-4 provides a United States Geological Survey geochronological timescale. Eons, eras, and periods are shown.

PRECAMBRIAN

The *Archaen* and *Proterozoic* eons are also known as the *Precambrian* eon. Rocks and fossils from the Precambrian time are calculated to be between 4 billion and 600 million years old, respectively. These super-ancient times, following the original formation and cooling of the Earth along with the formation of mountains, oceans, and much of the original development of life, are the subject of a lot of theory and speculation.

The *Archaen* (early *Precambrian*) eon is the most ancient time division and considered by most scientists as the beginning of the time divisions. This was the time when diverse microbial life thrived in the primordial oceans. The atmosphere was still very much anaerobic (lacking in oxygen) from the belching of ancient volcanoes and the development of the original landmasses.

The *Proterozoic* (late *Precambrian*) eon is the more recent of the two times. It was during this chapter in the Earth's history that the earliest forms of single-celled plant and animal life, like protozoa, were thought to have developed.

Following these first ancient time divisions, the *Phanerozoic* eon was further divided into the *Paleozoic*, *Mesozoic*, and *Cenozoic* eras. Table 2-2 shows the time divisions of different *eras* and smaller subdivisions into *periods* and *epochs*.

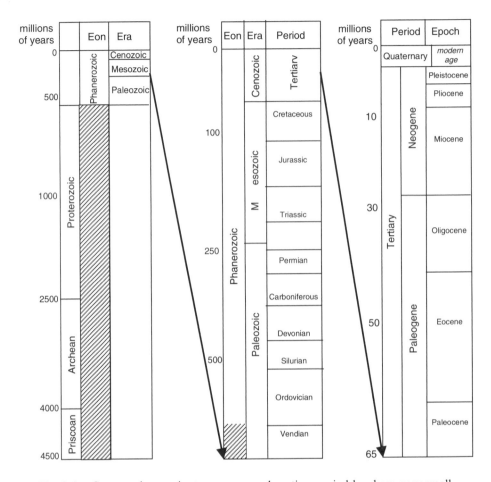

Fig. 2-4. Compared to ancient eras, our modern time period has been very small.

PALEOZOIC ERA

Roughly 300 million years ago, the oxygen in the atmosphere arrived at its present level. With its buddy the ozone layer to protect life from harmful ultraviolet radiation, the atmosphere of the planet allowed the development of life on land. This era was most favorable to the development and growth of *invertebrates* (spineless creatures like shrimp and jellyfish), fish, and reptiles. This generally tropical climate was divided by wide temperature

Table 2-2 Geological time is divided into eras, periods, and epochs.

Eon	Era	Period	Epoch	Years (millions)
Phanerozoic	*Cenozoic*	Quaternary	Holocene	0.1
			Pleistocene	2
		Tertiary	Pliocene	5
			Miocene	25
			Oligocene	37
			Eocene	58
			Paleocene	66
	Mesozoic	Cretaceous		140
		Jurassic		208
		Triassic		245
	Paleozoic	Permian		286
		Carboniferous		320
		Devonian		365
		Silurian		440
		Ordovician		500
		Cambrian		545

swings of different ice ages. By the end of the era, the continents were pushed up into the giant continent of Pangea.

As the landmass got drier, the humid swamps receded along with their unique plants and animals. This change caused the largest sweeping extinction of any of the eras. More life forms were lost than at any other time.

MESOZOIC ERA

This era can best be remembered as the era of the dinosaurs. Lasting only about half the time as the Paleozoic era, the Mesozoic era was a happening time. A time when plants, fish, shellfish, and especially reptiles were "super-sized," it was like everything on the Earth was on mega vitamins. Dinosaurs stomped around gigantic ferns and huge trees, while Pterosaurs (flying reptiles) cruised the skies. The climate was warm everywhere.

Although geologists can only guess at the forces that caused the break up of the supercontinent, Pangea, into Laurasia and Gondwana during this time, Antarctic samples hint at global "hot spots" of magma that built up causing cracks. Local dinosaurs and plants were separated for millions of years and became more specific depending on their own areas and the food and temperatures locally. Even small mammals began to pop up underfoot as chance appetizers for the meat-eating dinosaurs like *Tyrannosaurus Rex*.

During the Mesozoic era, more recent forms of insects, coral, ocean life, and flowering plants developed. All was really going great until suddenly, the dinosaurs and many other animals died off. Many scientists think this was caused by the impact of a large asteroid and the following years of global smoke, volcano eruptions, and generally nasty weather. The Sun couldn't shine through the ash and smoke, the water was contaminated, and the Earth was definitely not a great vacation spot.

CENOZOIC ERA

With the threat of the big dinosaurs gone, mammals thrived during the Cenozoic era. Early mammals lived pretty peacefully with birds, regular reptiles, and invertebrates. The climate became cooler and drier as the continents drifted apart and took on roughly their current positions. Some scientists suggest the Himalayas were pushed up during this time.

Perennial grasses grew and allowed the herds of grazing animals to flourish along with now-extinct side lines of the evolutionary tree. Temperatures continued to dive with the Antarctic landmass being formed. The appearance of the *homo sapiens* line of mammals appeared during the last few minutes of this era (geologically speaking) along with the use of primitive tools, discovery of fire, and the wheel along with extinction of more ancient species.

Look again at Fig. 2-4 for the portion of time in which our modern world has existed. If you were to draw the history of the Earth along a horizontal line, the few thousand years of known human existence would be little more than a sliver. Our modern, technological society of today would be less than a pin point.

When we talk about fossils in Chapter 10, we will see how the residents of the different eras responded to their changing environment.

Quiz

1. This era accommodated the development and growth of invertebrates
    (a) Paleozoic era
    (b) Cenozoic era
    (c) Mesozoic era
    (d) Jurassic period

2. With the threat of the big dinosaurs gone,
    (a) plants grew taller
    (b) mammals thrived
    (c) it rained more
    (d) large potholes became less of a problem

3. Scientists suggest the Himalayas were pushed up during which time?
    (a) Paleozoic era
    (b) Cenozoic era
    (c) Mesozoic era
    (d) Jurassic period

4. Clair Patterson and Friedrich Houtermans separately found the age of the Earth by
    (a) circumnavigating the globe in outrigger canoes
    (b) finding and comparing the radioactive decay rates of isotopes of copper
    (c) asking bright graduate students to help them
    (d) finding and comparing the radioactive decay rates of isotopes of lead

5. The supercontinent, Pangea, broke up into
    (a) Pangea prime and sub-Pangea
    (b) Laurasia and Gondwana
    (c) India and Canada
    (d) Antarctica and Holland

6. Geological time is measured in
    (a) meters/second
    (b) days/year

(c) geometric units

(d) geochronological units

7. Time that spans billions of years is known as
   (a) the beginning of the semester until the end
   (b) biological time
   (c) geological time
   (d) epidemiological time

8. Between Archaen and Proterozoic, which is earlier?
   (a) Proterozoic
   (b) Archaen
   (c) scientists are still observing the protists
   (d) they are the same age

9. Global "hot spots" of magma are thought to have caused
   (a) cracks in the super continent, Pangea
   (b) a thicker mantle
   (c) a change in vacation plans for early humans
   (d) a limited biosphere

10. Rocks and fossils from the Precambrian eon are thought to be as much as
    (a) 2 billion years old
    (b) 3 billion years old
    (c) 4 billion years old
    (d) 5 billion years old

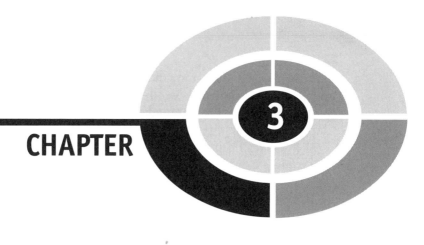

CHAPTER 3

On the Inside

Ancient people thought the Earth was flat and had no idea of its inner workings. Over time, they mapped out surface features like continents and oceans, but they pretty much kept with what was always known.

Europeans, who thought there was nothing to the east but more oceans, had to rethink that idea. When the Vikings and Christopher Columbus discovered whole new continental landmasses to overrun, it turned a lot of people's ideas upside down. They began to question all that they knew about the Earth, its creation, and ongoing development.

The stories of how the Earth was formed and what keeps it going have been varied over time and cultures. Around the globe, creative people came up with all kinds of explanations of what was going on in the center of the Earth. Early people thought that if they understood what made the Earth shake and destroy whole villages, especially if it was somehow their fault, they could prevent the problem from happening again.

Native Americans thought the Earth was like a mother and the sky a father that provided all their needs. This is easy to understand since everything, from their food supply (plants and herbs and buffalo) to their shelters and transportation (horses), were available from nature. When a natural disaster occurred, they took it as a sign that they had earned the displeasure of the earth or sky through careless stewardship of what they had been given. They thought they could prevent natural disasters from happening again by keeping a pure relationship with Mother Earth and Father Sky. Their problem with nonnative "invaders" was less about newcomers claiming the land for themselves (or their king), than their fear that any dishonoring of the land would bring disaster upon them.

In China, people thought a dragon lived in the center of the Earth. They believed that occasionally the dragon awakened, unleashed rumbling, volcanic spewing, and major disasters on the surface. Earth dwellers were at the mercy of the dragon and its fits of displeasure. By living in harmony with the Earth, people thought they would have a lot less chance of disturbing the dragon. Maybe this was the beginning of the expression, "Let sleeping dragons lie."

In India, people believed that the Earth was supported by elephants that stood on the back of a turtle that rested on a cobra. When the cobra moved, the turtle and elephants were unbalanced and the Earth shook and jolted in wild ways.

In the Polynesian islands, there are two legends describing the forming of the continents. The first describes how the gods fished the Earth out of the ocean, but the line broke releasing some of the catch back into the water. The part that stayed above the water level became the land and mountains. The second story tells how a turtle submerged deep in the ocean and came up with a huge clump of mud stuck to its back. This clump became the lands of the Earth.

Nebular Hypothesis

During the 18th century, early scientists moved beyond legends and started looking toward science for Earth Sciences answers. They started out by trying to understand how the solar system formed and where the Earth was placed in the system.

In 1755, Immanuel Kant, a German scientist/philosopher noted that the solar system must have formed from a large mass of gas and then gotten smaller and smaller from the tightening pull of gravity and rotation. After some time, rotation increased so much that the rings separated from the center mass. These rings eventually condensed to form planets that were held in orbit by gravity from the central mass.

Planetary Hypothesis

In 1900, Forest Moulton and Thomas Chamberlin, both from the University of Chicago, added their spin on this early theory. They thought that our Sun was a larger star before the planets were formed. A roving star with a strong gravitational field passed by and pulled a chunk of solar material away. This material spun off and in time condensed to form planets.

As we saw in Chapter 1, when the Earth was first formed, it was made up of a molten mass of simmering rock and metals. An outer cloud of elements that included hydrogen, helium, and carbon slowly circulated with heavy metals sinking deep into the molten core, and lighter elements rising to the surface. In this way, a multilayer crust was formed. This thin crust floated on a sea of molten rock for about four billion years, spitting volcanic gases.

Gravity had a large effect on this early formation of a layered Earth. From the core with its dense elements to the atmosphere with its light elements, the Earth's vertical differentiation was decided by gravity. Figure 3-1 shows the approximate densities of the Earth's layers from the atmosphere to the core.

The movement of continental landmasses affects the horizontal configuration of the Earth, but the development of the core, mantle, and crust came about as a result of gravitational forces.

The first atmosphere had little or no free oxygen. It was not user friendly and made today's pollution problems look like child's play. Some scientists think it wasn't until after the first single-celled blue-green algae appeared on a

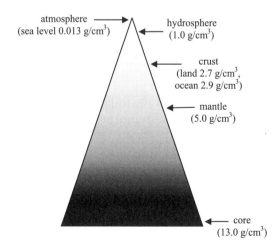

atmosphere
(sea level 0.013 g/cm^3)

hydrosphere
(1.0 g/cm^3)

crust
(land 2.7 g/cm^3,
ocean 2.9 g/cm^3)

mantle
(5.0 g/cm^3)

core
(13.0 g/cm^3)

Fig. 3-1.   Gravity had a big role in the vertical layering of the Earth.

global basis to metabolize toxic gases that breathable air was possible. Blue-green algae produce food by photosynthesis and as now, solar energy was plentiful. Oxygen is a side effect of the process. Remnants of these ancient organisms have been discovered in rocks over three billion years old.

After millions of years had passed, enough elemental oxygen was formed to provide an atmosphere for oxygen-breathing organisms. The Earth became a multilayered sphere of different temperatures and composition with a complex atmosphere and hydrosphere. Throughout this time, rain cooled the crust and collected in pools, rivers, lakes, seas, and oceans to provide good fishing, swimming, water skiing, and sailing millions of years later.

Crust

The crust of the Earth is a lot like the crust on bread, as far as the amount of crust compared to the rest of the loaf. The Earth's crust is just a thin skin on the land and under the oceans compared to the larger whole.

The outer layer of the Earth, or lithosphere, is made mostly of a brittle, rocky crust with the lower crust/upper mantle made of slightly less firm, but denser rock. The crust is where all the land that we know and love is found. It is the easiest layer to study by everyone, from school children bringing home shiny rocks to mom and dad, to petroleum geologists looking for the best places to drill for oil.

If the crust is made of rock, then what is *rock* exactly? Some people might call it a hard piece of dirt or soil, some might think it is a smaller part of a boulder, like a branch is a smaller part of a limb. These descriptions work for everyday, but geologists, who want to find out everything about how, when, and where the Earth's solid matter was formed, need to be more specific.

> To a geologist, a **rock** is an individual mass of solid matter that makes up part of the planet.

The key to the geologist's definition is that a rock is a mass of solid material. So then a handful of sand grains is not a rock because it is not cemented into a solid mass. If it was all one mass, like sandstone, then it would be considered a rock. A tree, though solid, is not a rock, but an ancient tree that has had all its organic material and water compressed out and replaced by minerals to become solid matter, is called *petrified wood* and is a rock.

The study of geology is about making simple observations. A lot of discoveries have been made by amateurs. But as instruments were developed that analyzed individual rock and mineral elements, as well as the Earth's vibrations, even more information was gathered.

The crust is also the thinnest of the Earth's layers making up only about 1/30th of the distance to the Earth's core. The top part is made of fairly light, granite-like rocks made of silica (SiO_2) and aluminum. These were formed from melted rock that pushed up from the mantle and other parts of the crust to become new land and mountain ranges above and below sea level.

The continental crust is made up of a mixture of rock types, mostly granites that are lighter in color and high in silica minerals. In fact, the Earth's crust is made up of over 70% silicon and oxygen. Table 3-1 shows the different elements found in the continental and oceanic crust.

Of the crust's minerals, the silicate group is the largest. It is based on silicon and oxygen with a mixture of different elements thrown in for color. Continental crustal rocks are made up of mostly granitic rocks, while oceanic crust is mostly basaltic rocks. Figure 3-2 gives you an idea of the amounts of minerals like calcium, silica, magnesium, potassium, and others found in the continental and oceanic crusts.

The underlying structure of all silicates is the crystal tetrahedron shape of silica. It is formed by a single silicon cation (Si^{4+}) bonded to four oxygen anions (O^{2-}). The different silicate mineral groups are separated by the way elements

Table 3-1   There is a variety of elements within the Earth's crust.

Element	Percent of the Earth's crust
Oxygen	47
Silica	28
Aluminum	8
Iron	5
Calcium	4
Sodium	3
Potassium	3
Magnesium	2

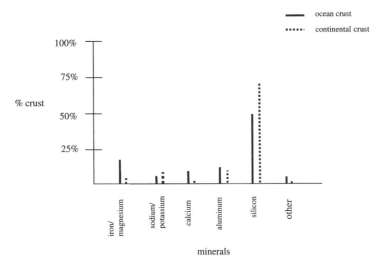

Fig. 3-2. Buoyancy causes a mirror image between upper and lower crustal elevations.

are bonded to the silicon tetrahedron. For example, the silicate *olivine*, a single-tetrahedron silicate, the anion charges of oxygen are balanced by the two positive cations of iron (Fe^{2+}) or magnesium (Mg^{2+}). The formula then is (Fe or Mg)$_2$SiO$_4$. Another common silicate, *garnet*, is balanced by calcium, magnesium, aluminum, or iron in the formula (Ca, Mg, or Fe)$_3$(Al or Fe)$_2$Si$_3$O$_{12}$.

The oceanic crust is heavier with more metals like magnesium and iron. The ages of the continental crust and oceanic crust are widely different. The continental crust is about 650 million years old, while the oceanic crust is only about 60 million years old.

Geologists suspect this huge gap of time is due to the more active recycling of the oceanic crust. The recycling of the crust is known as *subduction*. We will find out more about subduction and plate movement in Chapter 4.

The crust is made of big, broken chunks of land from the original super-continent, Pangea. This single giant landmass began breaking up about 200 million years ago. Figure 3-3 shows large land chunks (continents) scattered across the face of the Earth. So much of the Earth's crust has changed since its formation that deciphering the mystery of its growth and change will go on for a long time.

Mantle

The mantle, also known as the *mesosphere*, lies just beneath the crust. It forms over 83% of the Earth's volume and about 58% by mass.

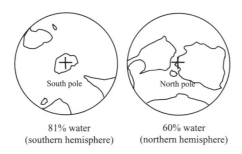

81% water
(southern hemisphere)

60% water
(northern hemisphere)

Fig. 3-3.   The layers of the Earth are so deep that geologists have only scratched the surface.

In the mantle, earthquake waves jump suddenly in speed (velocity) from the way they travel through the crust. This jump is determined by the change in density between the crustal rock and the denser mantle rock. This discontinuity between high and low velocities is called the *Mohorocičić discontinuity* or *Moho* for short. It is named after a Croatian seismologist and meteorologist, Andrija Mohorocičić, who first noticed its strange behavior after looking at seismic waves from the Kulpa Valley, Zagreb, Croatia earthquake in 1909.

Mohorocičić discovered that seismic waves came in two separate sets. Naming these wave sets, P- and S-waves, he noticed that one set arrived earlier than the other during the course of the quake. It was Mohorocičić's thought that one set had traveled through denser material than the other and was slowed by it. He proposed a theory that the Earth's outer rocky crust is about 30 km thick and rides on top of a denser mantle beneath it.

The **Mohorocičić discontinuity** is the boundary between the crust and the mantle.

When geologists began to track this wave change, they weren't sure what it meant and they tested a lot of different theories. As they gathered more and more measurements, they found that the speed of the P- and S-waves followed the variations in the thickness of the crust. This was seen in crust measurements from about 35 to 40 km below the continents and about 10 km below the oceans. However, below some high mountain ranges like the Andes, it can be as deep as 70 km in places.

This boundary between the crust and mantle became known as the *Mohorocičić discontinuity*. Research has fine tuned this slowing and found the seismic waves travel nearly 20% slower below the Moho than above it.

Most seismologists consider the Mohorocicic discontinuity, where it meets the upper mantle, to be the bottom or deepest limit of the Earth's crust.

Buoyancy

When geologists studied earthquake data further, they found that the crust was thinnest under the oceans and thickest below high mountain ranges. It was the thickest at the highest elevations and thinnest at low elevations. In other words, the Moho crustal boundary provides a mirror image of the crust above it. Figure 3-4 shows this mirroring of surface features.

This mirroring is based on buoyancy. The less-dense crust is floating on the pliable asthenosphere layer. Since buoyancy depends on thickness and density, the Moho boundary effect is a lot like that of an iceberg floating above the surface of the ocean. Icebergs float with only 10% of their volume showing above the waves. The density of water is 1.0, while the density of ice is 0.9 (because of the air trapped in the frozen water). The "tip of the iceberg"

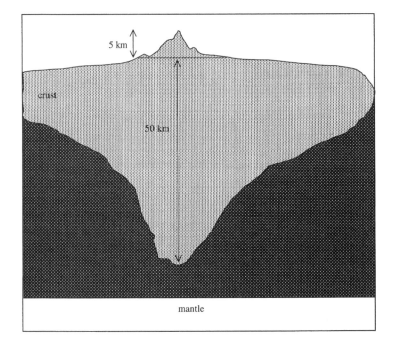

Fig. 3-4. Granite and basalt densities in the continental and oceanic crusts are different.

happens because ice is 10% less dense than water and 90% then, of the iceberg's volume is below the surface.

The continental crust is thicker under high mountain ranges to balance the floating "tip" above the land surface. Some geologists estimate that the depth to which a mountain's "foot" descends into the denser mantle is about $4\frac{1}{2}$ times the elevation of the mountain above. If this is true, then Mount Everest which stands about 8 km high must be supported by a crustal foot that reaches nearly 36 km, in addition to the 35 km of existing continental crust.

The oceanic crust is much thinner with few thickened spots. This is because it is made up mostly of mafic minerals that are heavy in iron and magnesium compared to the continental crust made up of mostly felsic minerals, richer in lighter, aluminum-bearing silicates.

Much of the Earth is made up of two pairs of elevations. One pairing is between 1000 m and sea level and the other pairing drops from sea level to 4000–5000 m below sea level. The first pairing includes the crustal continental platforms while the second pairing describes the abyssal oceanic plains. The balance between these two layers overlying the mantle allows density equilibrium to be maintained. The thickness gradient then allows for continental mountains and ocean basins. Ocean basins are low spots where water gathers, but flows across the lithosphere. Continental shelves create a gradual boundary into the oceans.

Temperature

Depth tests of mine shafts found that for every 60 feet drilled deeper into the Earth's surface, the temperature increased by one degree Fahrenheit.

$$1°\text{Fahrenheit} \uparrow /60 \text{ feet} \downarrow \text{ in depth}$$

The deepest shafts drilled into the Earth have been to a depth of about 13 km, but this is just a tiny prick compared to the total depth of the mantle. The entire mantle of the Earth is about 2897 kilometers (1800 miles) thick. Figure 3-5 gives an idea of the size comparison of different layers of the Earth and their incredible depths.

> **Core samples** are rock layer samples taken by drilling or *boring* at different depths of the mantle and bringing long cylinders of rock.

Bore holes and *core samples* are important in other ways. They give us information on the layering of the mantle as well as its makeup. Samples can

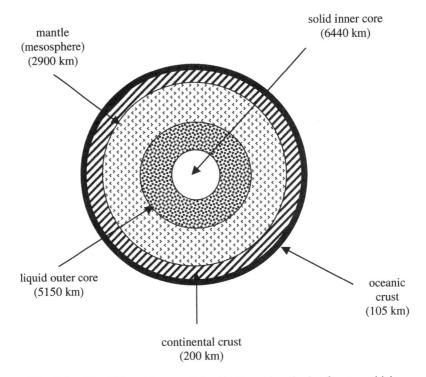

mantle
(mesosphere)
(2900 km)

solid inner core
(6440 km)

liquid outer core
(5150 km)

oceanic
crust
(105 km)

continental crust
(200 km)

Fig. 3-5.   The different layers of the Earth are hundreds of meters thick.

be analyzed for their content and percentages of different elements. Just like a core sample from a tree, a rock core sample shows different growth (or sedimentation) and composition patterns.

Electrical instruments that measure conductivity can also be used to take a look at the electrical properties of different layers of core samples. A sonic generator can be eased into a bore hole to provide a sound source to measure acoustic variations. Other sensors can be used to detect naturally occurring radioactivity levels of different elements in the layers of the crust. A combination of research tools are used individually and in combination with others to tell the overall picture of an area's geological profile.

For every kilometer drilled into the Earth, the temperature increases along a thermal gradient between 15 and 75°C depending on location. Just as the temperatures at the center of the Earth are extreme, the pressures are equally as intense. Temperatures have been estimated to be as great as 6000°C, with crushing pressures of 300 million kilonewtons per square meter or about three million atmospheres within the core. The size of the Earth allows a huge amount of energy to be stored within it as heat. The original heat of the planet is maintained by the constantly produced transformation, generation, and release of energy from radioactive elements.

Anyone traveling to the Earth's core would require vehicles found in science fiction that could withstand the intense heat and pressure. Otherwise, they would be fried and flattened like pancakes, not a good end for someone wanting to satisfy their scientific curiosity.

Core

Geologists cannot collect core samples and study the Earth's interior directly, so much of their information has been gathered from observations and clues from other sources. When the Earth's magnetism is measured, a variety of readings at different locations around the globe show a mixture of mass types within the planet.

The composition of meteorites gives scientists even more clues to the inner earth. These chunks of original matter from which the galaxy and solar system were formed continue to fall from space. Most burn up in the atmosphere because of the intense heat and friction, but a few larger chunks make it to Earth in one piece. There are two main types of meteorites: stony meteorites and iron meteorites. The stony meteorites are a lot like the mantle of the Earth, while the iron meteorites are more like the core of the planet.

The way *seismic* waves travel through the Earth are probably used the most to figure out how the Earth is put together to its core.

Seismic tremors or waves are made or related to the vibrations of the Earth. They are caused by earthquakes and other activities going on in the Earth's interior.

Geologists report that seismic waves show a major change in the way they travel and the material they travel through at a depth of 2900 km (1800 miles). The sudden shift points to the fact that the makeup of the Earth's inside changes at that depth. This is called the core–mantle boundary. Think of it like a peach with an outer skin, the fleshy fruit, and the woody pit. The fruit and the pit don't slowly morph into each other. The fruit doesn't gradually get tougher and harder until you reach the center, but changes abruptly from soft to hard.

But scientists don't have to wait around for an earthquake to test seismic activity, they can produce seismic waves with explosions or large vibrating machines on carrier vehicles. Then when the explosions or vibrations are begun, they measure the shock waves with special recording equipment called *geophones*, and then analyze the waves with computers to give complex pictures of how the wave patterns act. Results show how shock waves bounce off different layers within the crust and give geologists an idea of what a particular layer might be made of. For example, the speed of the waves would reveal whether a layer was solid or molten.

Seismic waves are known to travel slower through liquid than solid matter. Just as it is harder to drag your hand through water, compared to air, seismic waves go more slowly when traveling through liquid rock, compared to solid rock.

Using this knowledge, scientists found that seismic waves slow down when passing through the outer core, but speed back up when passing through the inner core. In fact, waves that don't normally pass through liquids at all are also blocked by the outer layer of the core. So scientists became fairly sure that the outer core is liquid or *molten*, rather than solid.

Molten rock is found at the innermost core of the Earth and in "hot spots" around the globe where internal pressures force it to the surface.

Waves change in strength according to the distance from their source and the types of matter they pass through. Seismic wave strength and behavior show density, movement, location, fluidity, and boundaries of different Earth layers.

Further evidence of a molten outer core is gathered from temperature readings. Miners found that rocks buried below the surface in the deepest shafts were hotter than those nearer the surface. When tested, shaft temperatures increased as depth increased. This excited scientists who had been puzzling over volcanoes, hot springs, and other geothermal sites for centuries.

We will study earthquakes more thoroughly in Chapter 12, but data shows that the core is made up of two major parts, the outer core, thought to be liquid, and the inner core, thought to be solid. The solid inner core is thought to be roughly 85% iron with small amounts of nickel, silicon, and cobalt. No one knows for sure how the Earth's core is layered because there is no way to drill to the center of the planet, but scientists continue to investigate with seismic testing. We saw in Table 3-1 that there are many different elements present in the continental versus oceanic crust.

Scientific Method

Just as Nicolaus Steno in the late 1700s began describing his three laws of relative dating based on his observations that we learned about in Chapter 2, so other scientists wanted to solve the Earth's mysteries. They wanted to understand all the workings of the Earth and to do this, they used the *scientific method.*

The keys to the scientific method are curiosity and determination, observation and analysis, measurement and conclusion. As humans, we are curious by nature. Throughout this book, you will get to know the tools and techniques that earth scientists use to find out all they can about our big, blue planet.

First, they start with a *hypothesis* like "fire burns" or "the Earth is flat." Then, they write down their observations about their hypothesis like "when fish is cooked too long over a fire, it turns to charcoal" or "on a calm day, the ocean is flat all the way to the horizon." People might follow a hypothesis for hundreds of years believing it to be absolutely true and then one day notice a new observation that presents doubt or completely proves it wrong.

> A **hypothesis** is a statement or idea that describes or attempts to explain observable information.

More and more information is gathered about the hypothesis like "all wood burns in a fire" or "fields of grain stretch for miles out upon the flat plain." Scientists then and now check to see if their observations always fit in different locations. Are all fires hot? Are all deserts flat? How do the tallest mountains fit into the picture? What are volcanoes and why are they hot?

> A **theory** comes about through careful testing and confirmation of a hypothesis over time.

Following years of testing by many scientists, experimental data is gathered that either supports a theory or blows it apart. (Always awkward when some segment of the scientific community wants to defend a favorite theory to the end.)

A theory predicts the outcome of new testing based on past experimental results. When a theory is found to be untrue, like when someone notices that the Earth's shadow on the Moon during eclipse is curved, not straight as a flat-edged ruler, then more observations and testing must be done to see if a new theory is needed.

To explain the differences in the size, shape, moisture, and composition of the landforms around the world, scientists presented the following theories:

1. Contraction theory,
2. Expansion theory,
3. Convection theory, and
4. Combination theory.

The *contraction theory* is the simplest theory stating that when the Earth originally cooled from the molten state, it became wrinkled and cooled unequally. The areas of hot and cool rock caused stresses that pulled, pushed, and compressed the land into different forms.

The *expansion theory* is roughly the reverse of the contraction theory. This theory suggests that the Earth was originally one-half of its present diameter of 12,756 km. It states that when the Earth cooled unequally at the beginning, there was an expansion of the faults and that gigantic blocks in the crust (about the size of Texas) were shoved up and out from the molten core. This led to increased volcanic eruptions and growing expansion of land from the seas.

In the *convection theory*, the upswelling of the land materials is thought to have happened through the circulation of *magma* or melted rock. As it expanded and pushed upward, it lost heat. The nearer the surface it got, the more it cooled and the denser it became. It turned into hard rock. When this happened, the hardened magma began to sink back down where it was heated, melted, and eventually started rising again. Scientists felt this theory explained the constant heat from the mantle and the increasing pressures.

The *combination theory* takes a bit from all the other theories and puts it together. In combination theory, the source of the energy that drives all this movement is thought to be the decay of naturally occurring radioactive elements supplying constant energy for the steady heating of the magma. In the combination theory, the crust is described as expanding along fracture lines, shoving broken blocks of crust apart, and making the pressure increase. Mountains are thought to be formed from a combination of melted rock and the folding and grinding of the landmasses sliding along on the "flexible" mantle.

In addition to the movement of the landmasses, scientists found another clue to support their theories. The composition of land rock and ocean rock was different. It was discovered that much of the landmass was made up of granite, while oceans contained mostly basalt. We will learn a lot more about

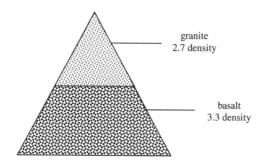

Fig. 3-6. Granite and basalt have different densities.

these types and others in Chapters 6–8 when we look at igneous, metamorphic, and sedimentary rock, but for now, just know that the two have different densities. Figure 3-6 shows the densities of granite and basalt. The movement of landforms is smooth when there is a lot of space. It can shift abruptly from pressure buildup after more land is packed into a tighter space. If you put a few grocery items on top of a carton of eggs in a paper bag, the eggs will be fine in their box. However, if more and more heavy items are piled into the bag, the pressure on the top of the egg carton increases and the eggs will break.

> A **law** is a hypothesis or theory that is tested time after time with the same resulting data and thought to be without exception.

When the same results are obtained over and over by a variety of experimenters, like "fire is always hot and burns in camp sites, farms, villages, cities, laboratories, and everywhere," then a theory is proven as a *law*. Over many years and repeated testing, laws are thought to be set in stone. If a new theory is developed, information is added to or takes the place of old ideas, and the cycle begins again. It takes a lot of testing and discussion for it to become a law.

In order for scientists to consider changing an accepted law, dozens of new experiments must be repeated by scientists all over the world that show the same results. Only this repetition of the same results will convince most scientists that a long believed law should be reexamined and might need to be revised or changed completely.

An example of this occurred in the area of medicine. Doctors for hundreds of years thought ulcers (sore spots in the stomach's lining) were caused by too much stomach acid. Antacids and diet change that limited acid production were common treatments.

Table 3-2 The dynamic Earth provides many different areas of focus for earth scientists.

Earth Sciences field	Area of interest
Astrogeology	Composition of the Earth compared to other planets
Astronomy	Location of Earth within the Universe
Cosmogony	Origin of the Universe and the formation of the Earth
Environmental geology	Conservation of resources and future planning
Exploration geophysics	Crustal composition to find resources (e.g., oil, gold)
Geochemistry	Chemical composition of rocks and their changes
Geochronology	Time as it relates to the Earth's history
Geomorphology	Nature, origin, development, and surface of land forms
Geophysics	Earth's magnetism, gravity, electrical properties, radioactivity
Glaciology	Formation, movement, and makeup of current glaciers
Hydrology	Composition and flow of water over the Earth
Micropaleontology	Microscopic fossils of plant and animal remains in rock
Mineralogy	Natural and synthetic minerals with a crystalline structure
Oceanography	Water makeup, currents, boundaries, topography, marine life
Paleontology	Identify, describe, classify, and date fossils
Petrology	Rocks
Seismology	Force, direction, and duration of earthquakes
Stratigraphy	Analysis of rocks and placing them in geological time order
Structural geology	Rock changes and distortions within the Earth's layers
Volcanology	Formation, activity, temperature, and explosions of volcanoes

However, in 1981, Dr. L. Robin Warren, a pathologist at the Royal Perth Hospital in Western Australia, found that in the stomachs of patients with ulcers, there were large numbers of the bacterium, *Helicobacter pylori*. When these ulcer patients were treated with antibiotics, they soon recovered. It took a long time for the medical community to accept this simple explanation for an illness that doctors had been struggling to cure for many years. It wasn't until a lot of other doctors and researchers came up with the same results that resistance faded and the standard treatment (law) for most ulcers became a cycle of antibiotics.

The keys to the scientific method are curiosity and determination, observation and analysis, measurement and conclusion. Humans are curious by nature. We question and study everything around us. The earliest people survived by trial and error. They kept what worked and didn't kill them, like foods and medicinal herbs, and avoided the things that did. In the following chapters, you will learn how earth scientists study the many faces and temperaments of this planet as well as what it might do in the future.

The study of Earth Sciences includes many different areas. Scientists look at different aspects of the Earth from the beginning of time to the present. If someone tells you they are an earth scientist, be sure to ask about their specific field of study. There is a world of possibilities! Earth scientists try to solve ancient and current mysteries of rock, atmosphere, oceans, glaciers, fossils, gems and minerals, earthquakes, volcanoes, and everything in between. Think of it like, "so many rocks (volcanoes, glaciers...you name the topic), so little time..."

Table 3-2 lists a few of the many fields earth scientists, in countries all over the world, are working in and attempting to decipher the Earth's internal and external processes.

As geologists better understand the Earth's rhythms and inner workings, theories and laws will continue to be strengthened or changed to reflect the latest experimental data.

Quiz

1. A law
   (a) is something invented by politicians
   (b) is a hypothesis or theory that is tested time after time with the same resulting data and thought to be without exception
   (c) cannot be changed
   (d) comes from observations made only once

2. Which theory is thought to be explained by the circulation of magma?
   (a) Convection theory
   (b) Combustion theory
   (c) Combination theory
   (d) Convolution theory

3. Which of the Earth's layers is the thinnest?
   (a) core
   (b) crust
   (c) edge
   (d) mantle

4. Magma is another name for
   (a) a small magnetic rock
   (b) magnesium
   (c) melted rock
   (d) arctic rock

5. In 1755, Immanuel Kant thought that the
   (a) Earth was 95% water
   (b) solar system was expanding
   (c) Moon was made from cheese
   (d) solar system was formed from a large mass of gas that shrank from the tightening pull of gravity and rotation

6. Scientists felt this theory explained the constant heat from the mantle and the increasing pressures
   (a) Convection theory
   (b) Combustion theory
   (c) Combination theory
   (d) Convolution theory

7. Granite is the rock found mostly
   (a) in aquariums
   (b) on the landmass
   (c) in the oceans
   (d) on the Moon

8. "The Earth is flat" is an example of
   (a) a limited perspective
   (b) ancient sailors' wisdom
   (c) a hypothesis
   (d) all of the above

9. The boundary between the crust and the mantle is called the
    (a) Mohorocičić discontinuity
    (b) Earth's limb
    (c) hot foot line
    (d) McKenna continuity

10. Bore holes and core samples
    (a) are a lot heavier when you bring them down from a mountain
    (b) are used to study the hydrosphere
    (c) give us important information about the layering of the mantle
    (d) provide little mineral composition data to geologists

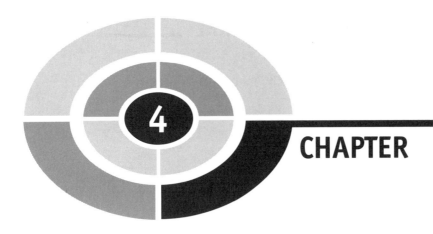

Plate Tectonics

Have you ever looked out of the window of an airplane and seen the widely different shapes, colors, and textures of the land below? Depending on the plane's height, the ground appears as an intricate carpet of every shade and texture. From rich browns and blacks to yellow, red, and every shade of green, the Earth's landmass is a mosaic of color. If you travel from the Midwest to the Northwest you will see everything from long stretches of grasslands with their circles of irrigated crops to salty deserts (ancient seas) of the west to the beautiful mountains formed from the clash of the North American plate with the Pacific plate. And that is just in the United States!

Travel across Africa, Australia, or Asia and see the variety and combinations of surface features there. Our world is a geological "gold mine" of diversity and mysteries waiting to be uncovered. The wide selection of lands and oceans seems beyond definition. New elements, mineral forms, and mechanisms are described nearly every year. It is an exciting field of study with tools that include things like axes, picks, compasses, and camping tents. It's a lot more fun than the standard bench top laboratory work. Remember, in Earth Sciences, the world is your lab; jeans and t-shirt your lab clothes!

Continental Drift

In the mid-1600s, Francis Bacon noticed that there seemed to be an odd, almost puzzle piece fit between Africa and South America. Not knowing what that meant, he put it down to remarkable coincidence. He told a lot of people about his theory and they thought it was interesting too, but nothing more than that.

Since the *Principle of Uniformity*, the idea that the Earth's past is based on its current form was firmly accepted, everyone figured that since the land-masses were anchored now, they must have always been anchored in about the same spots.

In 1858, after comparing the coastlines of Africa and South America closely, Antonio Snider-Pellegrini published his ideas of how the world looked before and after Africa and South America were pulled apart. Some people thought it was possible, while others rolled their eyes at the idea.

It wasn't until 1915, that Alfred Wegener, professor of meteorology and geophysics, in Graz, Austria, suggested the Africa/South America fit was the result of *continental drift*.

Wegener knew that scientists studying the fossil record of certain plants and animals were finding fossils in a narrow strip that stretched across several continents. Wegener was also studying the changes in world climate over time. He became aware of reports that fossils of plants that once grew in humid, hot climates were being discovered in polar areas and that fossils from colder climates had been found in hot, equatorial climates.

For example, *Glossopteris*, is an ancient fossilized plant found in southern Africa, Australia, South America, India, and Antarctica. Its huge pattern of dispersion seemed impossible to explain. Although some paleontologists thought the wind might have carried the seeds a long way, Wegener had a different idea. He thought the major continents had all been joined together in one piece at the time *Glossopteris* lived. Then, after the crust broke up and pulled apart, the continents drifted long distances from the places where *Glossopteris* first grew and then later turned into fossils.

Wegener decided that fossil rocks from climates we know today as having cold conditions were formed when their early land location was next to a geographical pole. He thought this was true even though some were now positioned at the equator.

Wegener studied many similarities of different landforms and was convinced that the original supercontinent, Pangea, developed many crossways fractures and drifted apart about 200 million years ago. To describe this in more detail, he published the *Origin of Continents and Oceans*. In it Wegener described how Africa and South America must have first split during the Cretaceous period, while much later, during the Quaternary period, Europe and North America, as well as South America and Antarctica, broke apart. He thought that even later in the Eocene period, Australia and Antarctica separated. Wegener could not explain what caused the original continental breakup and died while on an expedition to the Greenland icecap, but is considered to be the *Father of Continental Drift.*

One of Wegener's supporters, American scientist F. B. Taylor, published a paper in 1910, describing ancient movement of the mountain ranges in Asia from north to south. He thought mountain building was a lot more than just some in-place adjustments of the crust. He also thought the Mid-Atlantic Ridge was a crack that remained from the first pulling apart of Africa and South America. His ideas seemed to fit so well with Wegener's that some geologists called the whole splitting and drifting idea, the *Taylor–Wegener theory.*

In 1924, Swiss Alps and tectonics expert, Émile Argand spoke before the International Geological Congress in Brussels. In his talk he explained how it might be possible that the entire Alpine system, the mountains from the western Alps to the Himalayas, were formed from the drift of the Gondwana continent against Eurasia. He invented the word, *mobilism*, to explain sideways crustal movements and formation of mountain ranges.

A symposium (a meeting of experts) on continental drift was held by the American Association of Petroleum Geologists in 1926. Although the main organizers of the meeting were in favor of the continental drift idea, there were a lot of heated arguments about the existing data and what it meant. (That happens sometimes when you get a lot of scientists together!) By the end of the meeting, since no one could prove how continental drift took place, the majority of people didn't think continental drift was correct.

Finally, in 1928, Scottish geologist and Professor at the University of Edinburgh, Arthur Holmes, came up with the idea of an "engine" that might be providing the energy source for continental drift. Some people thought volcanic activity was the answer to continental drift, but Holmes didn't think so. He thought that a much higher energy source was needed to release the amount of heat produced by radioactive elements in the deepest layers of the Earth. He suggested that *convection currents*, the constantly circulating movements of heat and magma in the deepest layers of the Earth, provided enough energy to power continental drift.

Holmes knew that granites that make up a lot of the continents are high in radioactive elements. So he hypothesized that the temperature beneath the continents was probably higher than the temperature under the oceans.

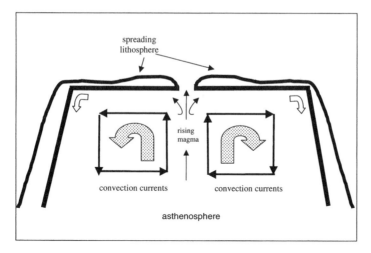

Fig. 4-1. Thermal convection currents play a main part in the Earth's magma movement.

If this was true, then convection currents would rise under the continents and spread out horizontally toward the continent's edges. After reaching the continent/ocean edge and lower temperatures, the cooler currents would turn and sink back downward.

Holmes knew that when a liquid was heated from below, the temperature increases upward until a critical temperature gradient is formed. When the temperature is increased even higher, the gradient is interrupted and thermal circulation currents form.

Have you ever seen a "lava lamp?" Picture the lazy blob movement in lava lamps. The waxy "lava" is heated by a bulb in the base of the lamp and when it gets hot enough, it rises upward. After reaching the top of the lamp, the lava starts to cool and eventually sinks downward again. When it gets back to the bottom and the hot light bulb, the lava heats and the whole cycle begins again. Figure 4-1 gives you an idea of how the Earth's thermal convection currents work. The expanding, newly formed lithosphere is also shown.

Plates

So what is a plate anyway? Sounds like something set out on the dinner table. In the study of the Earth, the science of geomorphology is connected with the study of landforms, all the bumps and grooves on the surface of the Earth. Several continental *plates* make up these landforms.

Geologists know that the original supercontinent broke up into large pieces of land like the size of the North American or African landmasses. These lumps of land are called *plates.*

How were land and ocean plates discovered? Well, there were lots of good clues to plate locations, since collision and grinding cause a lot of pressure buildup between plates.

Volcanoes that occur along plate edges offer active and dramatic fireworks to signal plate margins. Earthquakes and eruptions are concentrated along the boundaries of rigid lithospheric plates. The plates rimming the Pacific Ocean have so many active volcanoes that the area is known as the *Ring of Fire.*

> A geological **plate** is a layer of rock that drifts slowly over the supporting, upper mantle layer (asthenosphere) below it.

Continental and ocean plates are huge. They range in size between half a million to about 97 million km in area. Plates can be as much as 200 miles thick under the continents and beneath the ocean basins. Plates as much as 100 km thick fit loosely together in a mosaic of constantly pushing and shoving landforms. At active continental plate margins, land plates ram against other continental plates causing rock to pile up into towering mountains. Table 4-1 lists some of the mountain peaks found in the United States.

The border between the Eurasian and Indian-Australian plates is a good example of where plates clash. Along this plate margin, the Himalayan range is forming with the world's tallest mountain (Mount Everest). Where the Nazca ocean plate and South American continental plate collide, the Andes Mountains are forming. Similarly, where two ocean plates collide, one dives downward beneath the other and deep ocean trenches are formed. Like two stubborn bulls, the margin where the Pacific and Philippine plates meet created the Mariana trench (over 5 times as deep as the Grand Canyon).

All together, there are 15–20 major plates that make up the jigsaw puzzle of the Earth's crust. Of these, geologists consider that a few are small, some are medium sized, and several are massive. The 15 medium and massive plates are the most commonly studied plates. Figure 4-2 shows a United States Geological Survey illustration of the major oceanic and continental plates. Some of these plates are divided differently depending on the latest geology information, this gives a general idea of the main plates and their size.

The plates found across the face of the Earth are unique to this planet. If plates were thicker, they would surround the core like a pressure cooker until the temperatures and pressures became so extreme as to melt everything. This

Table 4-1   The Western United States has most of the highest peaks in the country.

Highest point	State	Elevation (meters)	Elevation (feet)
Mount McKinley	Alaska	6194	20,320
Mount Whitney	California	4418	14,494
Mount Rainier	Washington	4392	14,410
Mount Ebert	Colorado	4399	14,433
Gannett Peak	Wyoming	4207	13,804
Mount Mauna Kea	Hawaii	4205	13,796
Kings Peak	Utah	4123	13,528
Wheeler Peak	New Mexico	4011	13,161
Boundary Peak	Nevada	4006	13,143
Borah Peak	Idaho	3859	12,662
Mount Hood	Oregon	3426	11,239
Guadalupe Peak	Texas	2667	8749
Harney Peak	South Dakota	2207	7242
Mitchell	North Carolina	2037	6684
Clingmans Dome	Tennessee	2025	6643
Mount Rogers	Virginia	1746	5729
Mount Marcy	New York	1629	5344
Mount Katahdin	Maine	1606	5268
Black Mesa	Oklahoma	1516	4973
Brasstown Bald	Georgia	1458	4784

(*continued*)

Table 4-1   Continued.

Highest Point	State	Elevation (meters)	Elevation (feet)
Mount Sessafras	South Carolina	1085	3560
Mount Greylock	Massachusetts	1064	3491
Mount Davis	Pennsylvania	979	3213
Mount Frissell	Connecticut	725	2380
Timms Hill	Wisconsin	595	1951
Charles Mound	Illinois	376	1235
Jerimoth Hill	Rhode Island	248	812
Mount Woodall	Mississippi	246	806
Mount Driskill	Louisiana	161	535
Walton county	Florida	105	345

is the kind of thing that went on in the early forming of the planet. But the Earth is unique in that the story did not end there. It continued to cool, change, regenerate, and develop with a beauty and perfection that leaves earth scientists scratching their heads and most people of the world in open-mouthed wonder.

The main evidence that continents were originally all one piece comes from the discovery of rock formations that match from one continent to another. For example, the eastern edge of South America fits like a puzzle piece into the western border of Africa. Figure 4-3 shows how these two continents could have originally fit together. Fossils found on the once connected edges of North America, Europe and northwestern Africa all match.

Plate Tectonics

Plate tectonic geologists are always chasing their work!

Though they are always changing in size, the Earth has seven major continental plates. The outer crustal layer, the lithosphere, is a puzzle of

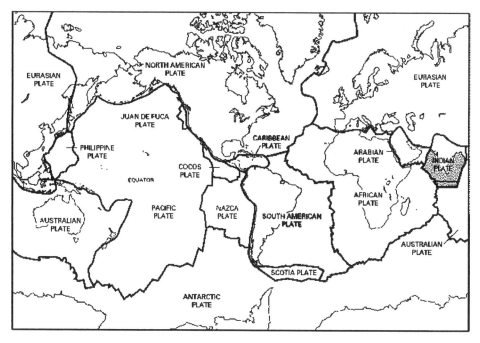

Fig. 4-2.   The Earth's lithosphere is made of many large moving plates.

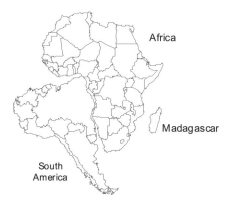

Fig. 4-3.   South America and Africa were probably one big landmass.

moveable parts that mold to each other according to the different pressures put on them. For the last two or three hundred years scientists have studied mountains, valleys, volcanoes, islands, earthquakes, and many other geologic happenings, but each study was done independently. Each individual study was thought to be unique and not connected to other geological sites or processes. Classification of rocks and land types was done apart from the other land types. It was not until widespread travel and communication began that geologists began to compare notes.

Geophysicist, J. Tuzo Wilson, was the first scientist to put it all together. He knew that *tectonics*, the large-scale movement and folding of the Earth's outer layers were ongoing. What Wilson pieced together from the ideas of Wegener and others was the concept of *plate tectonics*; the study of geology and physics. The little understood idea of continental drift made sense to Wilson when combined with the idea of large plate movement and pressures.

> **Plate tectonics** (*tektonikos* is Greek for "builder") describe the formation and movement of ocean and continental plates.

Plate tectonics is the umbrella theory that explains the Earth's activity and the creation, movement, contact, and flattening of the solid rock plates of the lithosphere.

But Wilson didn't just sit around his office thinking about plate movement, he led expeditions to remote areas of Canada and was the first to climb Mount Hague in Montana in 1935. The majestic range of mountains in Antarctica, the Wilson range, was named for this inventive and adventurous man.

Plate Movement

Since Wilson's first push toward the idea of plate tectonics, geologists began matching up plate measurements and found that plates moved farther over and around the planet than first thought. Most plates aren't even close to their original positions! Fossils of tropical plants, once located at the equator, have been found in Antarctica. Deep rock in the Sahara desert, sliced by the heavy passage of glacier travel, was frozen and frosty long before traveling to its hot and dry retreat of today.

Most importantly, plates continue to move, sliding along at rates of up to eight inches per year in some areas. Measurements made around the active Pacific plate shows lots of overall movement. The Pacific and Nazca plates are separating as fast as 16 cm/year, while the Australian continental plate is moving northward at a rate of nearly 11 cm/year.

Plates seem to move more slowly in the Atlantic where plates crawl along at 1–2 cm/year. Since the time of the first European explorers westward in the 15th century, the Atlantic plate has expanded by about 10 m.

Plates are affected most often by the movement of magma filling the cracks in mid-ocean floor ridges as the plates move apart. When this happens, magma pours out creating new ocean floor and edging along the existing plate margin. Across the oceans, there is an arrangement of ridges where new material is being formed. When enough new material is deposited, plates slant, slide, collide, and push over, under, and alongside their neighbors. Continental and ocean plates ride over or dive under each other, forcing movement down and back into the mantle and liquid core.

The regular arguing and conflict between plates causes and releases pressure buildup deep within the crust. Plate borders, sites of the highest volcanic and tectonic activity, are well known for their violent personalities. Ask anyone living in southern California, where the Pacific and the North American plates collide, about their many earthquakes and the Earth's constant rumblings!

> A **subduction zone** is an area where two lithospheric plates collide and one plate is forced under the other into the mantle.

Figure 4-4 illustrates the subduction of the lithosphere between plates. The lithospheric plate sometimes induces volcanism on the overriding plate. A crustal plate that is subducted then dives deep into the mantle. Note: Mountains and lithosphere not to scale.

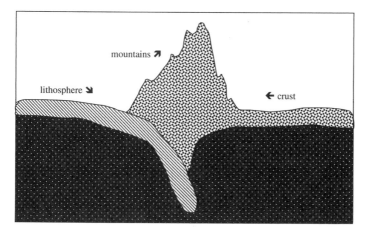

Fig. 4-4. Mountain building occurs at subduction zones between plate margins.

Convection

The circulation of material caused by heat is called *convection*. In the Earth system, convection is affected by gravitational forces within the planet as well as heat and radioactive recycling of elements in the molten core.

All tectonic processes within the Earth involve movement of solid or malleable matter. Convection in the mantle, driven by the thermal gradient between the core and lithosphere, takes place by deformation (creep) of the rocks and minerals that comprise the upper/lower mantle and the transition zone. Think of it like those square, hand-held puzzle games where one piece is left out and you can only slide one square into the open place at a time. In order to complete a number sequence or picture, you must keep sliding the squares around (one-at-a-time) until you are able to slide all of them around to their correct spots to complete the puzzle.

Mantle creep is like that. Because of imperfections in the crystalline structures of minerals and rocks, there are gaps. When pressure is applied, the atoms in the structure shift (creep), one atom at-a-time to a new position.

Plate tectonics, as seen in mountain building, earthquakes, and volcanoes, takes place by plastic (malleable) or brittle bending of the rocks and minerals that make up the oceanic and continental lithosphere. Temperature, pressure, and rate of deformation to a large extent define the nature of deformation for most minerals and rocks in the interior of the Earth. However, the chemical environment (presence or absence of water, oxygen, silica, and other elements) may also have a big impact. By understanding the mechanisms by which rocks and minerals move and change shape under extreme temperature and pressures, we will add to our understanding of the processes that shape our planet.

The steady movement of magma deep within the Earth depends on differences in temperature and differences in density within large "pockets" of molten matter. Depending on conditions, magma rises in the pockets of hotter temperatures and falls in pockets of cooler temperatures. Since the Earth's center is still hot, this endless thermal activity keeps the tectonic process going.

On a smaller scale, convection happens in liquids or gases, like the swirling currents of a pot of boiling soup. In the depths of the Earth, convection moves flowing magma that is heated from below by the core and then pushed upward over time and cooled from above. This solid flow movement is much slower than the liquid flow we saw earlier. Remember the lava lamp?

Convection is the process of heat transfer that causes hot, less dense matter to rise and cool matter to sink.

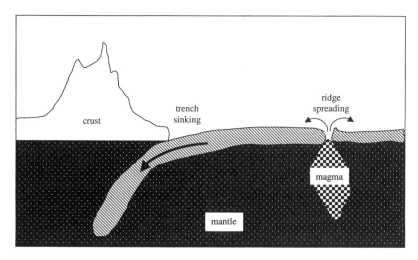

Fig. 4-5. Magma creates new land at ocean ridges.

Convection affects rocks of different densities as well. Lighter density lithospheric rock tends to ride along above sea level, while denser asthenospheric rock sinks below sea level. The hard, rigid lithosphere is an unyielding outer shell, while the softer, wax-like asthenosphere is moldable and fluid when pushed.

When hot matter is forced up and out, it cools and adds to the outer crustal rock. As more material moves up, the earlier matter is pushed out of the way. The pressure from underlying rock is removed as it comes to the surface and the material in the magma chamber, a "crystalline mush" heats as it gets closer to the surface. This activity expands the area between the plates by a few centimeters per year. After a while, this new surface rock comes to another plate that will not yield. When this happens, plates argue and new rock gets pushed back down by subduction to be melted over again. Subduction occurs between opposing plates (mountains and magma chamber are not to scale), while cooling, rising magma causes spreading at ocean ridges as is shown in Fig. 4-5.

Plate Boundaries

Convection causes plates to meet and separate. When this happens, there are three main types of plate boundaries that form. These boundaries include:

- *Convergent boundaries* – plates clash and one is forced below the other, pulling older lithosphere to the depths of the mantle (see trenches below).

- *Divergent boundaries* – plates pull apart and move in opposite directions making room for new lithosphere to form at the lip from outpouring magma (see ridges below).
- *Transform fault boundaries* – plates slide past each other parallel to their shared boundaries.

Since the Earth's core is hot in the extreme, roughly 6000°C, the malleable mantle beneath the brittle crust is always hot too. This constant heat production by the core keeps the cycle of convection going. The heat transfer from rising and sinking convection currents provides the power that moves plates around the globe.

The rate of plate movement varies a lot depending on location. In Africa, there is very little movement from year to year, while the active Pacific plate has sections that move as much as 10 cm/year relative to the hot spots.

TRENCHES

A long, thin ocean valley, sometimes less than 100 km wide, with steep sides and caused by the descent of a plate's edge back into the mantle, is called an ocean *trench*. Some of the deepest points on Earth are found within ocean trenches. The Java trench in the West Indies and the Mariana trench in the Pacific average between 7450 m and 11,200 m.

A trench is formed along the *convergent boundary* of two plates. Subduction digs ocean trenches when one plate collides with another, pushing it down underneath the first and causing a deep trench. The front edge of the top plate is crumbled and pushed up like snow in front of a snow plow. The clashing forces and constant pushing action along the border between two plates, form towering mountain ranges parallel to the trench like the Andes range along the Peru–Chile trench.

Before the idea of global plate tectonics was accepted, marine geologists were stumped over the formation of ocean trenches. They didn't understand what was causing the ever deepening valleys in the ocean floor. They kept trying to figure out why the core or lower mantle seemed to be pulling down the asthenosphere. They didn't know much about convection currents at that point and so had no energy source for the movement of the landmasses.

Because most subduction zones are found in the Pacific Ocean, the edges of the Pacific plate, where surface rock is constantly being pulled down and destroyed, has the most deeply grooved trenches. The Pacific Ocean is ringed by these trenches because of the constant plate action of the Pacific oceanic plates against the North American, Eurasian, Indian-Australian, Philippine, and Antarctic plates.

Trenches are found at both continental margins and at ocean–ocean convergence zones along island plate lines. The Java trench, also known as the Sunda trench is a deep depression in the Indian Ocean, 305 km from the coasts of the islands of Sumatra and Java, Indonesia. The trench is 2600 km long and is the deepest point in the Indian Ocean.

Twenty-two trenches have been identified though not all are major trenches. Of these, 18 are in the Atlantic and one (Java trench) is in the Indian Ocean. The depths of the major trenches are greater than 5.5 km deep and between 16 and 35 km in width. The deepest trench is the Challenger Deep (11 km deep) found in the Marianas trench. The Peru–Chile trench, off the coast of South America, is the longest trench at 1609 km in length, while the Japan trench at 241 km is the shortest.

RIDGES

A *rift* or upgrowth of the ocean floor, where plates are slowly edged apart by the filling of hot magma, is known as an ocean *ridge*. Ridges are formed along *divergent boundaries* where plates move slowly away from each other. Magma then rises into the crack between them, filling it, and hardening into rock. Figure 4-6 shows how this seafloor growth takes place.

Most of the magma exiting the mantle today is found at ridges in the ocean floor and along plate edges. When magma pours out of cracks in the ocean floor, they build up a lip along the crack and form mid-ocean ridges. Ridges thousands of miles long can be found in the Atlantic Ocean, and around the plate borders of the Pacific plate.

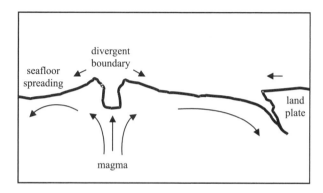

Fig. 4-6.   Upswelling magma adds to seafloor spreading at divergent boundaries.

Cooled magma (lava) also flows horizontally forming more ocean floor and piling up around vertical *vents* to form volcanic cones and islands like the Hawaiian Islands and the Galapagos chain. These are hot spots. The unending creation of solidified magma (rock) creates new seafloor and widens the ocean basins, a process called *seafloor spreading*.

When British geologists, Drummond Matthews and Fred Vine sampled rocks along the edges of ocean ridges, they found that the farther away they were from the ridge crest, the older the rocks. When this information was added to the idea of continental drift and seafloor spreading, it helped explain the puzzling increase of crustal landmass and supported the plate tectonics theory.

Nearly all of the ocean ridges are at the bottom of the oceans, but the Mid-Atlantic Ridge that stretches up the center of the Atlantic Ocean, emerges in a few places including Iceland, where geologists can measure its growth and characteristics.

Below the waves, photographs from submarines at great ocean depths show that rocks near ridge edges are clean and sharp. As the distance from the ridge increased, rocks became covered with sediment. At about 10 km (6 miles) from a ridge, the rocks are completely obscured from sight by layer upon layer of sediment dusted over them for millions of years. We will take a closer look at the hardening of sediment into rock in Chapter 6.

Transform Fault Boundaries

A fault is simply an opening between two plates caused by plate pressure that builds up until the surrounding rock can't take it anymore and splits.

> A **fault** is a fracture or zone of fracture in the crust, where some type of movement happens.

Some plates don't clash head to head, but instead slide past each other horizontally in what is known as a *transform fault*. The rock on either side is moved in opposite directions as the buildup of pressure between the plates provides the energy for movement. Figure 4-7 shows the displacement of the boundary line in a transform fault.

Fault blocks or sections of rock on either side of the fault can be lifted up on one side or both. They can have faults on one or more sides and can be lifted up on one side and dug in on the other depending on the surrounding rock type.

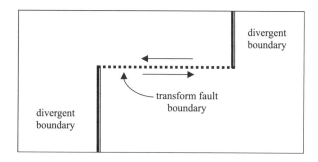

Fig. 4-7.   Transform fault displacement along two plates.

The well-known San Andreas Fault in California where the North American and Pacific plates meet is a *transform fault boundary*. Along this fault, the Pacific Ocean plate is sliding north while the continental plate is moving southward. Since these two plates have been at it for millions of years, the rocks facing each other on either side of the fault are of different types and ages. From the air, sharp differences in color and texture are obvious.

The "Big Bend" area of the San Andreas fault is responsible for a lot of the intricate faulting in southern California. A fault bend is often found in a confined area of plate collision. A tremendous amount of compressional pressure is created. To release this stress a bit, additional faults form over time. Commonly, *crustal shortening* happens as a response to intense compression. Crustal shortening allows compression to continue by packing rocks tighter in a compressional zone. When this occurs, shorter *thrust faults* are created.

> **Thrust faults** are the low-angle reverse faults that pack crust sections over one another to create a thicker mound of crust with a shorter (horizontal) length.

Not all the pressure generated by the bend of the San Andreas fault goes into thrust faults. The collision margin is at an angle, so that some of the in-between rock is able to move sideways out of the way. Large regions of sideways faulting have formed in order to relieve some of the stress created by the fault bend. Figure 4-8 shows how these horizontally formed faults are compressed around the area of the fault. We will discuss faults and their different types in greater detail when we study earthquakes in Chapter 12.

As with most plate collisions, transform faults do not slide along smoothly at a constant rate, but in fits and starts. Extreme grinding friction is caused

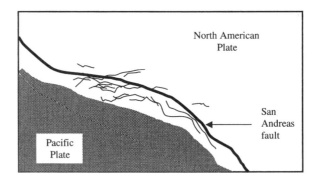

Fig. 4-8. Compression adds to the creation of short horizontal faults.

by the buildup of pressure between the two clashing plates. This pressure is usually released by earthquakes and a sideways slip between the transform fault fractures. The 1906 and 1989 earthquakes near San Francisco were caused by side-slip transform fault movement.

Following a slip, pressure builds up for many years until it again reaches a critical pressure point, like the straw that broke the camel's back. One day, when the "last straw" is added by pressure buildup from the mantle, everything shifts violently again. This sudden movement causes millions of dollars in damages to populated areas: breaking roads, building foundations, bridges, and gas lines. Fires are also common following earthquakes when gas, freed from lines broken during the side-slipping of two grinding plates, ignites.

IGNEOUS ACTIVITY

For millions of years, interior magma has bubbled up from the Earth's mantle only to cool, turn solid, and add to the depth of the crust. When held back for any length of time, the pressure increases to an extreme point until it blasts violently through volcanoes to form new rock along their sides. The comings and goings of magma is called *igneous activity*, from the Latin word, *ignis* or fire. Magma that erupts from a volcano is called *lava*, which when cool turns into *volcanic rock*. This rock name comes from the Roman God of fire, Vulcan. The study of volcanoes is sometimes called *vulcanism*. We will study volcanoes more completely in Chapter 11.

Rock from magma that bubbles up more slowly and never reaches the surface is called *plutonic rock*. It was named after the Greek God of the underworld, Pluto. Plutonic rock spends most, if not all, of its lifetime deep within the Earth.

CONTINENTAL SHIELDS

All continents are made of new and old rock. When Pangea fractured into several chunks and began drifting around the face of the globe, land originally side by side drifted hundreds and thousands of miles apart. It was this "sameness" of rock types, in far flung areas of the world, that got geologists thinking that all land must have been together in one piece originally.

Further study of these ancient areas showed that the lowest level of crustal rocks, known as *granulites*, formed a kernel around which the continents developed. These dome-shaped structures or *shields* have very little sedimentary deposits and only thin soils.

Ten to twelve continental shields have been discovered containing ancient rocks. The largest of these are the Canadian Shield in North America and the Fennoscandian Shield in northern Europe. The western one-third of Australia has been found to be part of an ancient shield.

Every continent has an area of ancient unchanged rock known as a *continental shield*. These stable, shield areas have experienced very little change. Since the original Precambrian eon formed continents millions of years ago, continental shields have only felt minor bending and gentle erosion compared to highly stressed plate margins. Surrounding the continental shields are flat, sediment containing *continental platforms*.

CRATONS

In addition to the continental shields, geologists find areas of rock that form an edge or frame along the rim of the shields. These edge areas are called *platforms*. When shields are framed by a platform area, it is known as a *craton*.

Cratons are made up of pieces of continents that have not been affected by major changes since Precambrian times.

The four billion-year-old metamorphosed granite, known as the Acasta Gneiss found in the Northwest Territories of Canada, shows that the first kernels of continental crust were around even during the earliest formation of the Earth.

When chunks of granitic crust combined into stable, solid kernels drifting about on the malleable mantle, they provided a place for cooled bits of rock to buildup. The first cratons formed about 1.5 billion years ago, with larger pieces the size of Australia and India and smaller bits the size of Madagascar. These early cratons drifted about on the upper mantle until they cooled and slowed down long enough to stick together in larger and larger masses. Eventually, they grew to become continental landmasses with pushed up mountain peaks and ranges.

The North American continent is made up of seven cratons that fused together millions of years ago. These combined craton landmasses account for about 80% of today's continental landmass with the ancient rock masses making up only a tiny part of the total landmass.

The Earth's constant magma recycling melted most of the first rock-forming kernels since their formation millions of years ago and transformed them into new rock over much of the planet's solid surface. Canada, Africa, and Australia are the only known places that still have rocks unchanged throughout geologic time. In the United States, the oldest Precambrian rock is found in the nearly two billion-year-old Vishnu Schist at the bottom of the Grand Canyon.

GREENSTONE

Metamorphosed lava and sediments from volcanic eruptions in the Pre-cambrian period were formed early on when the Earth's crust was warmer and more malleable. The large crustal plates still floated freely, and were added to by huge, violent explosions of lava that blew through cracks and holes in the new crust.

Rock formed in this way is known as *greenstone*. After ash and lavas bubble through seawater and groundwater of temperatures between 150 and 300°C, greenstones are formed. The typical green color comes from high amounts of chlorite. Greenstone rock contains most of the world's gold. Most gold mines around the world are ancient playgrounds of volcanic activity. Greenstone belts in southeast Africa are about 19 km thick and roughly three billion years old.

OPHIOLITES

Ancient ocean floor sediments that have turned to rock and pushed up through cracks in the continents are known as *ophiolites*. However, before plate tectonics was explained, geologists couldn't figure out how these rocks, usually found on the seafloor, came to be located on land.

It was not until samples gathered by submarines and deep-sea drilling were studied more closely that scientists figured out this mystery. They found that ophiolites are formed when the oceanic crust that has been smoothed and smashed against the continents is carried along with seafloor spreading and then shoved up onto the land. This process has been going on for a long time. Some of the ophiolites samples studied are thought to be around 3.6 billion years old.

Ophiolites, with veins of rich ores and mineral deposits, are found in many of the mountain ranges of the continents.

BLUE SCHISTS

Blue schists are metamorphosed rock of subducted oceanic crust forced back into the mantle at subduction areas of the ocean floor or forced up onto the continents. There are also green schists that contain larger amounts of chlorite and epidote and are formed by low-temperature, low-pressure metamorphosed volcanic rock. We will learn more about metamorphosed rock and its formation in Chapter 8.

MEASUREMENTS

With the development of sonar and highly sensitive imaging instruments during World Wars I and II, the timing of plate tectonics was right. The world's ocean depths were determined and ridges and trenches discovered. Plate motions could be drawn accurately with much less guesswork.

The use of the Global Positioning System (GPS) (the same system that allows some automobiles to know exactly where they are on a road trip) uses the radio signals of an encircling network of 27 GPS satellites, each with a highly precise atomic clock on board. A ground-based radio receiver gathers the signals from 4 to 7 satellites at the same time and identifies the differences in the movement time from each satellite. A component of the receiver uses the time differences to locate the receiver to within 1 cm. Along the San Andreas fault in southern California, there are nearly 300 GPS monitoring stations constantly checking satellite signals for small displacements in local landforms.

NASA's Space Shuttle and the International Space Station also provide valuable, real-time imaging. Through precise measurements, geologists have been able to accurately calculate the spreading of the Mid-Atlantic ridge to within a centimeter and the slow closing of the Pacific Ocean through subduction.

Plate movements are used by geologists to help to predict possible earthquakes and volcanic eruptions. This "early warning system" gives scientists around the world, one more way to protect entire populations from Mother Nature's occasional temper tantrums. We will look closer at some of these tantrums in more detail in later chapters.

Quiz

1. Who is called the father of continental drift?
    (a) Sherlock Holmes
    (b) Alfred Wegener
    (c) Francis Bacon
    (d) Alexander Fleming

2. Oceanic crust sediments and fragments that have been smoothed and smashed against the continents are called
    (a) amorites
    (b) trilobites
    (c) ophiolites
    (d) smashorites

3. Pieces of continents that have not been affected by major changes since Precambrian times are called
    (a) cratons
    (b) pylons
    (c) glaciers
    (d) blue schists

4. The area where two crustal plates collide, forcing one plate under the other into the mantle is known as the
    (a) Wegener zone
    (b) Subduction zone
    (c) Twilight zone
    (d) Benny and Jerry zone

5. The Canadian in North America and Fennoscandian in Europe are both examples of
    (a) balmy climates
    (b) continents in the southern hemisphere
    (c) types of bacon
    (d) shields

6.   Blue schists are formed as a result of
     (a) translation
     (b) subduction
     (c) a blue moon
     (d) convection

7.   An ocean ridge is formed when
     (a) currents have piled a lot of pollution of top of itself
     (b) glaciers retreat
     (c) continental shelves break off
     (d) ocean plates are edged apart by the infilling of hot magma from
         below

8.   Transform fault boundaries occur when
     (a) one plate is pushed under the next
     (b) plates slide past each other at roughly 0° angles to their shared
         boundaries
     (c) plates slide past each other at roughly 45° angles to their shared
         boundaries
     (d) plates collide and are unable to move in any direction

9.   Violent explosions of lava that blew through cracks and holes in the
     new crust formed
     (a) fossils
     (b) blue schists
     (c) greenstone
     (d) peridot

10. Who described the concept of plate tectonics in terms of geology and
     physics?
     (a) J. Tuzo Wilson
     (b) Jack Wilson
     (c) John Lennon
     (d) Antonio Wegener

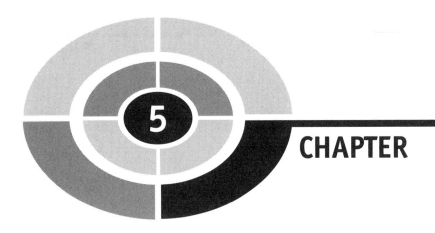

CHAPTER 5

Strata and Land Eras

Land eras are the broad ranges of time that geologists use to group different information. For example, if a geologist wants to talk about the time of the dinosaurs, the Jurassic period might be mentioned. Remember, *Jurassic Park*, the "science-gone-terribly-wrong" movie where scientists use genetic material preserved in petrified tree sap to produce prehistoric dinosaurs? It couldn't be named Cambrian Park because the much earlier Cambrian period was home to mostly microorganisms, ancient horseshoe crab-like invertebrates (*trilobites*), and other shelled inhabitants. They aren't nearly as exciting to watch on the big movie screen.

We will learn more about the specific types of sediments laid down by rivers, glaciers, ocean movement, decay of microorganisms, and other factors in Chapter 7, but first, let's step back and look at the bigger picture. Why are geologists interested in studying sedimentary layers in the first place? What kind of geological history can be discovered through studying different rock forms?

Well, geologists look at strata (rock layers) like pieces in a history puzzle. The make up, depth, type, angle, and compression of sedimentary rock give geologists an idea of the "how and when," of rock deposition.

> **Strata** is the layering of the Earth's sedimentary rock layers into beds, either singly or layers upon layers over geological time.

When visiting deep cuts into the Earth's sedimentary rock layer, like the Grand Canyon, millions of years of sedimentary layering can be seen. Geologists try to reconstruct the Earth's developmental history and formation by studying these rock layers. The theory of plate tectonics came into full acceptance after decades of careful measurements and study of the formation and movement of the Earth's strata by geologists. The study of the Earth's strata, known as *stratigraphy*, also allows the sequencing of formation events.

> **Stratigraphy** includes the formation, composition, sequence, and relationship of sedimentary rocks in strata.

Law of Original Horizontality

Most sediments were deposited beneath the seas and oceans of the world. Because of gravity and the more or less horizontal ocean floor, sediments are originally laid down in flat, horizontal layers. This nearly universal process has become known as the *Law of Original Horizontality*. These layers may get pushed, folded, erupted on, and other things, but they started out flat.

Principle of Stratigraphic Superposition

Have you ever been in an ancient city and seen evidence of many civilizations who built their idea of the latest architecture on top of older structures?

When the basements of towering, new office buildings are dug, often times an area's history is revealed as well. Even cross-sections of older city streets sometimes expose older and earlier layers of brick, cobblestones, and dirt beneath recent concrete paving.

When this happens with layers of sediment and sedimentary rock, it is called the *Principle of Stratigraphic Superposition*. This principle says that in any strata, which has not been folded or overturned, the oldest sediments will be found at the bottom of the sample with the most recent sediments layered on top.

> The **Principle of Stratigraphic Superposition** says that the deeper you go into the Earth, the older the sedimentary rock.

This law applies to sedimentary, igneous, and metamorphic rock layers, as well, that haven't been mangled by any other outside forces after their first layering.

The principle was first described by William Smith, a civil engineer who did a lot of surveying work for canal construction in western England. During the construction of the canals, he noticed that there were different layers of sedimentary rock in a predictable order of layering. In 1816, Smith published *Strata Identified by Organized Fossils* in which he listed 17 strata with specific plants and organisms unique to each for periods between Jurassic and Tertiary. The next year, he added 10 more strata downward until he reached bedrock granite.

Smith got so used to seeing the different rock layers that after a while, he could name the layer, the region it came from, and its position in the rock sequence. He was a regular rock detective! During Smith's study of sedimentary rock layers, he also found there were certain fossils that seemed to be connected to specific layers. This fact helped him identify the layers and their most common order of deposition. Later, this fossil and sediment relationship became known as the *Law of Faunal Succession*.

> The **Law of Faunal Succession** explains how fossil faunas and floras follow one another in a definite, identifiable order.

Geologists use the Law of Original Horizontality, the Principle of Stratigraphic Superposition, and Law of Faunal Succession to figure out the age, scattering, and order of different layers of sedimentary strata.

Sedimentary Facies

When a sequence of sedimentary rocks is examined from one layer to another, clear differences can be seen between the layers. These differences are based on the environmental conditions at the time they were deposited. For example, some geologists think the dinosaurs were killed off as a result of a huge meteor impact after having discovered a thin layer of dust and ash over much of the Earth. They think the impact caused fires and dust that rose high into the atmosphere and was suspended for years before finally settling back to the Earth as sediment. This ash and dust layer was compressed into sedimentary rock.

> **Sedimentary facies** are a common group of characteristics within a sedimentary layer (unit) that are specialized as a group.

Unique facies are used to interpret the depositional environment. As you move across a continent and then into the ocean, you'll notice a wide variety of environments with separate characteristics like grain size and shape, color, deposition, stratification, or fossils. Each new facie can be totally different or slowly change to take on a new composition and texture.

Lithofacies or rock facies are the rocks in a specific facies group. For example, one set of strata may be composed of a sandstone facies, a shale facies, and a limestone facies. Where rocks are not exposed to the surface, it is much more difficult to figure out what is going on. If a limestone rock is exposed in one area and a piece of granite in another, it is a lot harder to determine if the two are of the same age, same facies unit, or whether they were deposited at the same time.

Stratigraphic Record

The *stratigraphic record* is the overall picture of a series of facies in a region. Stratigraphy and structural studies of continental rocks have allowed geologists to piece together the principles and relationships of physical geology. Plate tectonics was finally figured out after decades of stratigraphic observations and careful measurements. Layer upon layer of sedimentary rock over millions of years and during all types of weather and environmental changes have provided geologists a road map of ancient times.

There are also breaks in the stratigraphic record. Since sedimentation happens at different rates all around the world, some places have a barely readable record. Wind, erosion, and other factors can wipe out sediment before it has a chance to gather, let alone harden into rock.

> **Unconformity** is a good-sized gap or break in the stratigraphic record that shows that a part of the rock record is missing.

When geologists see an unconformity in the stratigraphic record, they know that part of the story is missing. There are several factors that can break the sedimentary rock record. These include the uplifting of landmasses, changes in sea level, and changes in climate that change streams, rivers, and glaciers.

At the ocean floor, strong currents move sediments and can cause underwater landslides of deep trenches and other large areas where sediments pile up. Sometimes these changes cover up other strata making a geologist's job even trickier.

When there is a large break or gap in the rock record, geologists call the missing time period a *hiatus*.

> A long-term gap in the sedimentary rock strata, affecting thousands and millions of years, is known as a **hiatus**.

No solid information on sedimentation, climate, or tectonics is obtainable for the time period represented by the hiatus. Geologists have to "wing it" and make assumptions based on their experience and knowledge of the surrounding area's history.

Some unconformities are much shorter and temporary than those seen during a hiatus. These unconformities are the newsworthy events that television reporters and journalists look for. Sudden natural events are called *diastems*.

For example, when huge tropical storms wash away decades of beachfront habitats and collected sand, sedimentation is affected. When locally heavy rain swell rivers beyond their banks causing swift flooding and erosion for miles, it affects the crops that year, but not for the next one hundred or a thousand years.

> Brief gaps or strata disruptions, caused by sudden events like flash floods or mud slides, are called **diastems**.

There are four different types of unconformities. These are the *angular unconformity*, *disconformity*, *paraconformity*, and *nonconformity*. These missing puzzle pieces are found in a variety of different circumstances.

Figure 5-1 illustrates the different kinds of unconformities commonly found in the crust. The individual differences between the unconformities are described below:

- *angular unconformity* has a break between older and younger strata with the one forced upward at an angle to the other,
- *disconformity* has parallel strata layers with a rough surface break erosion,

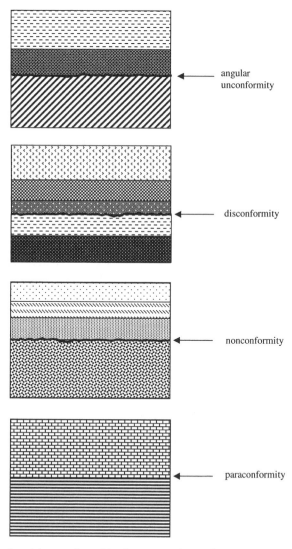

Fig. 5-1.   Unconformities are found in four types depending on angle and adjacent rock.

- *nonconformity* has stratified rocks that lie over igneous rock and metamorphic rocks with a rough break between them, and
- *paraconformity* has parallel beds divided by an unconformity of only a different bedding plane between them.

Stratigraphic Classification

The Earth's crust is constantly changing inside and out. Sedimentation and erosion processes are high for long periods (millions of years) and then something changes and they slow or stop. As we saw with plate tectonics, continental landmasses are shoving and compressing at their margins, while ever-widening ocean ridges spread apart. The crust is always on the move and brings about everything from nearly bottomless trenches and impossibly high mountains to crystalline beaches and black lava fields. Our "blue planet" is anything but boring.

Unlike newly erupted igneous surface rock, sedimentary rock strata give us snapshots of individual climates and geological events throughout history. When geologists put all these snapshots together, it creates a "family album" of all the species, habitats, landscapes, and temperatures of the Earth.

By studying layers of different kinds of rock, geologists get not only knowledge of strata composition, but also a window into the experiences and influences that were in place during a specific time period. Geologists pull all this mixed information together in a system of *stratigraphic classification* that can be used by scientists all over the world.

ROCK STRATIGRAPHY

The study of rock stratigraphy is basically a grouping exercise. It reminds me a lot of the sorting exercises we did as children. What belongs with what? Which of these things goes together and in what order? At first glance, the many layers of a sedimentary rock structure look like a crayon box full of different colors or an artist's box of paints. Differences are easily spotted between natural tones and earthen hues. But in addition to that, geologists have the added benefit of texture. Figure 5-2 shows a cross-section of the ancient (Precambrian) and more recent (Paleozoic) sedimentary rock layers that make up Arizona's Grand Canyon in the United States. Some layers are thin, some thick, some rocky, and some smooth, but all have a place in the geological stack. Each particular band added together makes up the

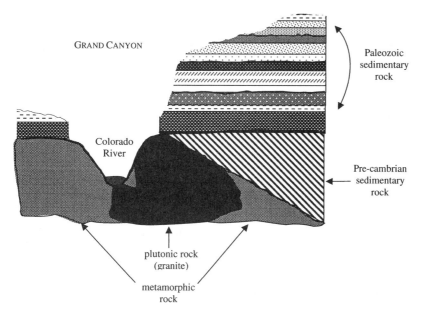

GRAND CANYON

Paleozoic
sedimentary
rock

Colorado
River

Pre-cambrian
sedimentary
rock

plutonic rock
(granite)

metamorphic
rock

Fig. 5-2. The Grand Canyon is a colorful stratigraphic record of sedimentary rock.

total vertical picture. Some layers are separated by unconformities. The Grand Canyon is a well-known example of stratigraphic sedimentary rock layers that lie above metamorphic and original plutonic rock.

> An individual band in vertical strata, with its own specific characteristics and position, is called a **rock-stratigraphic unit** or **rock unit**.

When several rock-stratigraphic units are stacked vertically, they add up to a *formation* which geologists can then describe and map as part of the geological record.

Formations are collections, then, of many rock-stratigraphic units grouped together into a section with the same physical properties. Formations are commonly thick enough to be seen in a lot of different places where various strata layers are exposed. Igneous and metamorphic rock layers also have specific formations. Two or more formations can also be bunched together into *groups*.

When drawing geological maps, different formations are called by name like the Green River formations. When naming a formation, geologists usually use the name of a surrounding area or the formation's major stone type, like the red sandstone formations in Red Rock Canyon, Nevada.

For even more detail, geologists subdivide the physical characteristics of formations into smaller rock-stratigraphic units called *members* and even smaller divisions, called *beds*. We will learn more about sedimentary rock beds in Chapter 7.

When studying sedimentary rock strata, even more than igneous or metamorphic rock, it is important to remember the huge stretches of time that have led to the layer upon layer of solidified rock. Thousands and millions of years have added atom upon atom, crystal upon crystal to slowly build each layer. It is kind of like watching paint dry multiplied a million times slower. In order to better understand the super slow deposition of sedimentary rock, geologists divide strata by periods of time called *time-stratigraphic units*. Then, when they are discussing a certain formation, they can further divide it into sandstone formed at one time, compared to nearly identical sandstone, formed much later.

> **Time-stratigraphic units** are the rock layers with known characteristics that formed during a specific period in geologic time.

Time-stratigraphic units are commonly based on the fossil groups they contain and are sorted to represent progressively shorter time periods. These major groupings are combined into *systems* and systems are combined into *erathems*. Smaller geologic time units are further divided into *eras*, *periods*, *epochs*, and *ages*. These help to further track changes in stratigraphic rock layers over time.

DIACHRONOUS BOUNDARIES

Sometimes the upper and lower margins in an area are slightly different from the main body of the sedimentary facies. This happens in an area where sedimentary deposition increases as more of a changing gradient than a one-time event.

> **Diachronous boundaries** grow at different rates both laterally and with respect to time.

In a river delta or marshlands area, for example, where some sections are dry, some marshy and some marine, the deposition rate of each area can be changed by the amount of local rainfall, temperature, and commercial activity. If a massive storm comes along one year and washes away a lot of

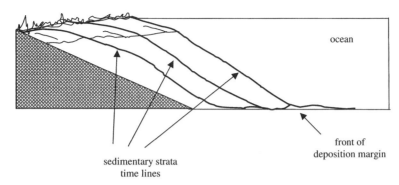

sedimentary strata
time lines

ocean

front of
deposition margin

Fig. 5-3.   Sedimentary margin lines show the progress of sedimentation in an area.

the silt collected in one area, then the silt replacement in the following years
will be younger than the untouched prestorm silt buildup.

In the same way, the increasing front edge of a silting river delta is younger
than layers below it or sections higher up the river.

Diachronous boundaries are found in areas where sedimentary rock
is laid down at different times in different areas.

When changing sedimentation occurs with diachronous boundaries and
then becomes rock, the related time lines can be traced from further up the
river, down the slope of the piled silt, to the delta's leading edge of buildup.
Figure 5-3 shows how this silting gradient might look.

STRATIGRAPHIC COLUMN

Although rocks are grouped together by time and composition, they can
also be grouped together into broad sequences of strata separated by
major unconformities. These unconformity-limited sequences are made
up of strata that have margins at the base and top by area and regional
unconformities.

Geologists use *seismic stratigraphy* to outline these sequences. Through the
use of high-resolution seismographs, the stratigraphic column can be finely
detailed in most of its layers. By getting a cross-section of crustal rocks and
sediments, seismic profiles of specific structure, thickness, regional environment, and unconformities can be drawn up.

Regional unconformities can be mapped across broad sedimentary rock
basins within a subcontinent. Sometimes they are pushed up and folded by

tectonic activity. Sedimentary rock solidification in a specific sequence can also be found stretching across a continent all the way to the continental shelf. By using seismic profiles, global spreading and layering of sediments can be followed from continent to continent to show how they might have originally looked when they were in larger land chunks.

> **Lithology** is the study of the physical characteristics of a rock through visual recording or with a low-power microscope or hand-held magnifying glass.

Lithologic similarity is the matching of different rock formations separated by small and great distances by their physical characteristics. These physical characteristics include the following specific traits:

- grain size,
- grain shape,
- grain orientation,
- mineral content,
- sedimentary structures,
- color, and
- weathering.

The main drawback of this type of comparison is when the distance is so great between formations that environment and weathering can make them look very different. The topography of an area rising above a lower plain can also make identification tough. Plants and trees add to the problem, but can be useful if viewed from the air. Geologists use changes in different plant growth as a clue to the soil. Some plants can't grow in sandy or rocky soils, so show different growth patterns according to their base soils. Geologists study these growth preferences to figure out where collected sediments start and stop.

Key beds

To help them follow lithographic changes, geologists use *key beds* as a marker to certain types of rock. This is particularly useful to geologists that live and work in a certain area for many years. The key beds are so familiar that they are almost like old friends.

> A **key bed** is a thin, broad bed with very specific characteristics that are easily seen and recognized.

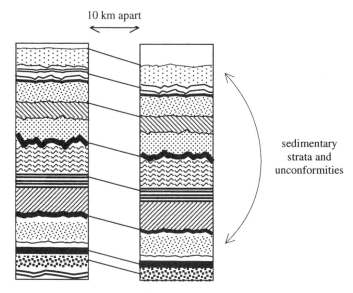

10 km apart

sedimentary
strata and
unconformities

Fig. 5-4. Stratigraphic columns have matching layers even at a distance.

Key beds can be tagged from one outcrop to the next, even miles apart, if their main characteristics are specific enough. Color can be a big player in key beds. A thin, black layer of volcanic ash, easily seen on top of pale tan silts or sandstone, can be used to correlate many key beds in a region.

It makes sense, then, that if separate key beds are able to be correlated, then the strata just above and below the key bed would correlate as well. This sameness of the *stratigraphic column* helps geologists match key beds, formations, deposits, and time periods. It is one of the main factors used in figuring out relative time connections between regions. Figure 5-4 shows key beds and layers between two stratigraphic columns found 10 km apart.

Electrical, radiometric, and fossil dating factors

Well samples taken from drilled holes in sedimentary petroleum-rich areas can be compared as well. Just as tree core samples have rings of growth, so too do rock core samples. Along with actual samples, energy companies also use instruments that can be lowered into drill holes to read *electrical properties* of the layers below. In this way, they read sedimentary shifting across distances and determine where to drill next.

Radiometric dating is used to get much closer to actual age when dating sedimentary rock samples. This type of dating, described in Chapter 2, is fairly direct except when other types of rock, like igneous rock, cut through the sedimentary rock layer being tested. It is important to remember that radiometric dating, like most mathematical calculations, have a standard error or deviation from the calculated number. Radiometric dating is thought to have between 5 and 20% standard error when dating rocks thought to be millions of years old.

Another type of dating factor used when comparing layers is a *biological factor*. Geologists look at the types of plants and microorganisms living in the sediment at the time it was deposited. For example, if an animal was found on the Earth only 500,000 years ago, it would be impossible (or at least very wrong) for a layer containing fossils of that animal to be dated at 80 million years ago.

The upper and lower boundaries of a formation are determined by their characteristics. These may be marked by unconformities or *guide fossils*; fossils repeatedly found to be associated with a certain period of time.

> **Guide fossils** are those wide-spread genus and species of fossilized organisms found within a specific rock-stratigraphic unit.

We will learn "everything you ever wanted to know about fossils in the Earth" when we look at their different sizes, types, developments, migrations, and fossilization in Chapter 10. Until then, remember that they are another piece of the puzzle geologists use to date sedimentary rock strata.

GEOLOGIC TIMESCALE

When geologists gather an area's data together to decipher the total geological picture, they consider the known key beds, strata, and formations in chronological order and include electrical, radiometric, and fossil information. This stratigraphic analysis helps them figure out the *geologic timescale*. This timescale is divided into four major eons that have been decided on the basis of the life that lived during that time.

> The four major divisions of the geologic timescale are the **Hadean, Archean, Proterozoic,** and **Phanerozoic** eons.

The first eon is known as the *Hadean* eon. Hadean, Greek for "beneath the Earth," is the earliest rock record. This most ancient rock is also found on other planets that don't have constant volcanic activity to change it from the original. The second eon is the *Archean* eon, a slightly less ancient eon. The third eon, the *Proterozoic*, is thought to have seen soft multicelled microorganisms, but much of this rock has been weathered away or changed. The fourth eon is the *Phanerozoic*. This last major eon contains hard-shelled microorganisms that are fossilized and studied today.

It is this last major eon, the Phanerozoic, that has been further divided into the eras described in Table 2-2. These are the Cenozoic, Mesozoic, and Paleozoic which are then further divided into the Quaternary, Tertiary, Cretaceous, Jurassic, Triassic, Permian, Pennsylvanian, Mississippian, Devonian, Silurian, Ordovician, and Cambrian periods, respectively. The Quaternary and Tertiary are further subdivided into seven epochs of the Holocene, Pleistocene (Quaternary) and Pliocene, Miocene, Oligocene, Eocene, and Paleocene (Tertiary). Since the time frame described is so long, it was easier to divide it up into manageable chunks.

The life forms found in these different eons and eras were individual enough to set them apart from the earlier ones. Periods and epochs, however, tend to have blurry boundaries.

When Charles Lydell came up with the idea of uniformitarianism, he based it on the fossil communities found in the sedimentary rock of samples taken in Italy and France. These references are important in the general concept of similar fossils being found together, but today we can find more variety around the world in sedimentary rock than what Lydell knew about at the time. As better methods of uncovering fossil and rock samples are developed, we will have even more information and may break down the timescale again.

STRATIGRAPHIC BOUNDARIES

Timescale subdivisions of sedimentary stratigraphic units are based on their community makeup. The microscopic inhabitants of strata and the way they change over time are very important in relating time periods to each other. Figuring out whether organisms came on the scene earlier or later than others, depending on development, is a question geologists ask. These organisms are found mostly at the sites of ancient seas and oceans, but land-based animals are also used to figure out timescale. *Stratigraphic boundary changes*, then, seem to be based on all inhabitants, land and sea, in a specific area.

When geologists find a major event that killed off, depleted, or shifted the majority of strata's inhabitants, they try to place it. These events are thought to be related to a catastrophic event like earthquakes, volcanic eruptions, or something equally sudden. When these catastrophes take place, along with constant tectonic activity, strata can have very different compositions.

Some geologists think stratigraphic boundary changes depend on global climate changes. These changes are thought to occur as a result of sea level rise and fall and tectonic clashes. When the oceans of the world are lower, like during times of heavy glaciation, the exposed land forms can be eroded by wind and rain. Erosion in an area reduces and may even eliminate known strata. Later (thousands to millions of years), when a warming in global temperature melts frozen ice, the seas rise, land is covered again, and sediments are deposited again. Figure 5-5 shows the amount of water covering the landmasses about 80 million years ago.

Study of the fossil strata of continental margins has found times when the seas covered much more of the land than they do now. The early Paleozoic Era and Cretaceous Period show a thick layer of marine sedimentary rock on nearly every continent during this time. The stratigraphic record experiences erosion of the sedimentary rock only during times of the fall in sea level. When these same locations are sampled, a buildup of sediment during times when the water level increased again is seen.

As plate tectonics shove and push continents around, the shape of ocean basins is changed. Sometimes the land crustal mass is increased and sometimes the ocean crust is increased in area. This depends on subduction and other tectonic forces.

Think of it like building a sand castle at the beach. Depending on the castle's shape and surrounding moats, distance from the water's edge and tide level, the details of the feudal kingdom will remain for a long time or will be quickly washed away.

Other experts believe a huge global impact from space caused the sedimentary deposition of a thin clay layer seen in many parts of the world. This clay layer, called the "boundary clay," has been found to contain high

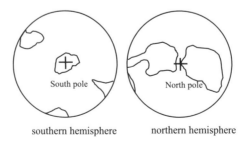

southern hemisphere   northern hemisphere

Fig. 5-5. Ancient seas covered much of the landmass long ago.

levels of iridium and platinum elements. Usually found in very low levels of different crustal rocks, this clay layer is thought to have come from the great amounts of rock and dust thrown into the air upon impact. The incinerated forests and vegetation would have caused fires and smoke that blocked out sunlight for months to years. Photosynthesis would have been blocked, killing off plants that provide food for larger creatures in the food chain. In addition, a sharp drop in solar heating would have thrown the Earth into a cold period. The cold would have intensified ice formation and encouraged an ice age.

Sedimentary rock samples taken in the western United States show that plant populations were radically changed at the end of the Cretaceous Period, with some species becoming extinct at the iridium rich clay layer. Then the next sedimentary layer shows fern growth, a plant that pops up following major ecological change.

Geologists have supported this comet/meteor impact theory with calculations that an object, roughly 10 km in diameter, impacted the Earth at the time of the Cretaceous/Tertiary boundary. Statistical research has shown that an impact once every 100 million years is likely. For the Phanerozoic eon (570 million years), five major extinctions have been studied or about one every 100 million years.

However, some geologists think this same boundary clay could have been caused by a period of intense volcanic activity. Something like all the volcanoes of the world going off at the same time. It's hard to tell what happened, but the layer exists. It is perhaps best used as a time-stratigraphic marker in the geologic column.

When correlating strata based on physical and biological information, there are many different characteristics to consider. A geologist trying to decipher the thickening and thinning of sedimentary rock over time looks at many different geologic sites and samples.

The main thing to remember is that the Earth holds lots of clues to its past. We don't have to puzzle over our past empty handed. For the person who likes mysteries, geology holds a lifetime of fun.

Quiz

1. The deeper you go into the Earth, the older the sedimentary rock is called the
   (a) principle of paleontological facies
   (b) principle of stratigraphic superposition

(c) principle of growing older
(d) paleotectonic mapping technique

2. Sedimentary facies are
    (a) a type of gemstone used in machinery
    (b) made up of rock fragments erupted from volcanoes
    (c) a common group of characteristics within a sedimentary layer
    (d) the only type of rock found in the Grand Canyon

3. Which of the following is not a type of unconformity?
    (a) transitional unconformity
    (b) disconformity
    (c) paraconformity
    (d) nonconformity

4. Stratigraphic units are commonly
    (a) rock and water
    (b) rock and time
    (c) air and time
    (d) water and time

5. The fact that sediments are normally laid down in flat, horizontal layers is called the
    (a) law of perpendicular positioning
    (b) law of minimal friction
    (c) law of sedimentary consolidation
    (d) law of horizontal originality

6. Strata is
    (a) a type of waltz
    (b) the type of coloration seen on a zebra's coat
    (c) defined as unchanged metamorphic rock in the mantle
    (d) the layering of the Earth's sedimentary rock layers into beds over time

7. Stratigraphy includes all of the following sedimentary rock factors in strata except
    (a) temperature
    (b) sequence
    (c) relationship
    (d) composition

8. All of the following are geological time units except
    (a) periods
    (b) eras
    (c) commas
    (d) epochs

9. Which period is the Holocene epoch found in?
    (a) Tertiary
    (b) Jurassic
    (c) Quaternary
    (d) Triassic

10. The visual study of a rock's physical characteristics is known as
    (a) cryptology
    (b) lithology
    (c) paleobiology
    (d) paleontology

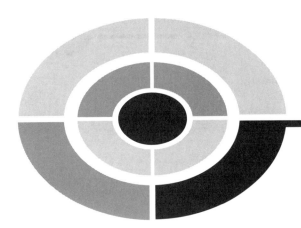

Part One Test

1. Past locations of the magnetic poles and the Earth's crustal movement can be found from
   (a) watching polar bears
   (b) the study of rocks with magnetic particles
   (c) watching which direction water swirls down a drain
   (d) ice cube shift in a glass of soda

2. Mount Everest
   (a) the ultimate challenge for mountain climbers
   (b) has low levels of oxygen near its peak
   (c) is the tallest peak on the Earth
   (d) all of the above

3. Archean, proterozoic, and phanerozoic are
   (a) types of ancient lizards
   (b) eras in the late Cenozoic
   (c) types of eoliths
   (d) three major eon divisions

4. Which of the following is not a type of unconformity?
   (a) paraconformity
   (b) angular unconformity
   (c) osteoconformity
   (d) disconformity

5. How many feet into the crust must be drilled before a 1° Fahrenheit increase in temperature is seen?
   (a) 30 feet
   (b) 40 feet
   (c) 50 feet
   (d) 60 feet

6. What is the deepest trench on Earth?
   (a) Andes
   (b) Samoan
   (c) Mariana
   (d) Himalayan

7. Who came up with the idea of how South America and Africa looked before they were pulled apart?
   (a) Arthur Holmes
   (b) Galileo Galilei
   (c) Francis Bacon
   (d) Antonio Snider-Pellegrini

8. Areas of rock that form an edge or frame along the rim of the shields are called
   (a) platforms
   (b) shield enigma
   (c) sediments
   (d) platesh

9. The word Archean comes from the Greek word for
   (a) archer
   (b) ancestor
   (c) ancient
   (d) anchovie

10. Scientists found that the composition of land versus ocean rock was
    (a) the same
    (b) different

(c) unequal

(d) not important since most of it was below the waterline

11. Flattened, dead volcanoes along the East Pacific Ridge are called
(a) guyouts
(b) gutters
(c) fissures
(d) voltimeters

12. Which era was the most favorable to the development of spineless creatures like shrimp and jellyfish?
(a) Mesozoic
(b) Embryonic
(c) Cenozoic
(d) Paleozoic

13. The formation, composition, sequence and relationship of sedimentary rocks in strata is called
(a) paleontology
(b) stratigraphy
(c) metamorphology
(d) cartography

14. Ignis is the Latin word for
(a) cat
(b) yellow
(c) sand
(d) fire

15. Fossilized ants and insects are often found intact in a petrified tree sap called
(a) hematite
(b) turquoise
(c) amber
(d) olivine

16. Increased volcanic eruptions and the growing expansion of land from the seas is part of what theory?
(a) Contraction theory
(b) Adhesion theory
(c) Convection theory
(d) Expansion theory

17. Roughly how many plates make up the Earth's crust?
     (a) 7
     (b) 15
     (c) 35
     (d) 45

18. The mantle makes up approximately what % of the Earth's volume?
     (a) 50%
     (b) 72%
     (c) 80%
     (d) 97%

19. Mobilism is the word Émile Argand used
     (a) to explain horizontal crust movements and the formation of mountain ranges
     (b) for the study of large white whales
     (c) for several layers of sedimentary rock
     (d) to describe ejected ash and stone

20. The Mesozoic era is best remembered as the
     (a) molten era
     (b) conservative era
     (c) dinosaur era
     (d) industrial era

21. All of the following are geological time units except
     (a) colons
     (b) eras
     (c) epochs
     (d) periods

22. Hadean, Archean, Proterozoic, and Phanerozoic are
     (a) types of cod fish
     (b) major divisions of the geologic timescale
     (c) geologic epochs
     (d) architectural forms

23. Which Scottish geologist proposed an "engine" that could provide an energy source for continental drift?
     (a) Arthur McDonald
     (b) Arthur Holmes
     (c) Terry Pappas
     (d) Elisabeth Holmes

24. What is the basic process that makes "lava lamps" work?
    (a) gravity
    (b) circadian rhythm
    (c) solar energy
    (d) thermal convection

25. Dating samples is
    (a) fairly simple if you wear the right outfit
    (b) not an exact science
    (c) a lot like dating geologists
    (d) an exact science

26. Magnetometry is
    (a) a game of three players and a magnet
    (b) the measurement of the Earth's rotation around the Sun
    (c) a variation of geometry
    (d) the measurement of the Earth's magnetic field

27. A thin, broad bed with very specific characteristics that are easily seen and recognized is called a
    (a) twin bed
    (b) sofa bed
    (c) key bed
    (d) day bed

28. Roughly how many centimeters per year does the Atlantic plate move?
    (a) 1–2 cm
    (b) 3–4 cm
    (c) 5–6 cm
    (d) 7–8 cm

29. A zone, where two crustal plates collide and one plate is forced under the other into the mantle, is known as a
    (a) isolation zone
    (b) subduction zone
    (c) conduction zone
    (d) convection zone

30. The heated materials of the asthenosphere become less dense and rise, while
    (a) cirrus clouds bring in extra moisture
    (b) the core continues to cool
    (c) cooler material sinks
    (d) mountain peaks shift southward

31. The era when plants, fish, shellfish, and especially reptiles were "super-sized," was called the
    (a) Platonic era
    (b) Mesozoic era
    (c) Cenozoic era
    (d) Hypnotic era

32. Molten rock is also called
    (a) extremely hot
    (b) sediment
    (c) amber
    (d) magma

33. Which period is the Eocene epoch found in?
    (a) Tertiary
    (b) Jurassic
    (c) Quaternary
    (d) Triassic

34. The core makes up what percentage of the total mass of the Earth?
    (a) 30%
    (b) 40%
    (c) 50%
    (d) 60%

35. Eras are divided into smaller subdivisions called
    (a) periods and epochs
    (b) subcontinents and land masses
    (c) epochs and endochs
    (d) periods and courses

36. The Andes mountains were formed when which two plates collided?
    (a) Juan de Fuca and Pacific plates
    (b) African and African plates
    (c) Nazca and South American plates
    (d) Eurasian and Philippine plates

37. Wide-spread fossilized organisms found within a specific rock-stratigraphic unit are known as
    (a) a fossil clump
    (b) fish fossils
    (c) trilobites
    (d) guide fossils

38. Tectonics comes from the Greek word (*tektonikos*) for
    (a) builder
    (b) technician
    (c) skyscraper
    (d) water

39. The hydrosphere, crust, and atmosphere all make up the
    (a) magma
    (b) core
    (c) biosphere
    (d) ionosphere

40. Some scientists suggest the Himalayas were pushed up during this era
    (a) Paleozoic
    (b) Mesozoic
    (c) Crytozoic
    (d) Cenozoic era

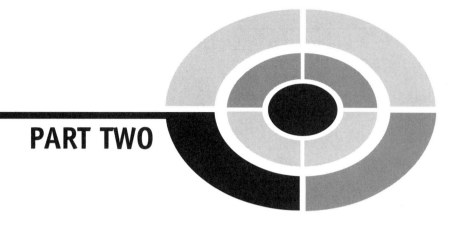

Minerals and Rocks

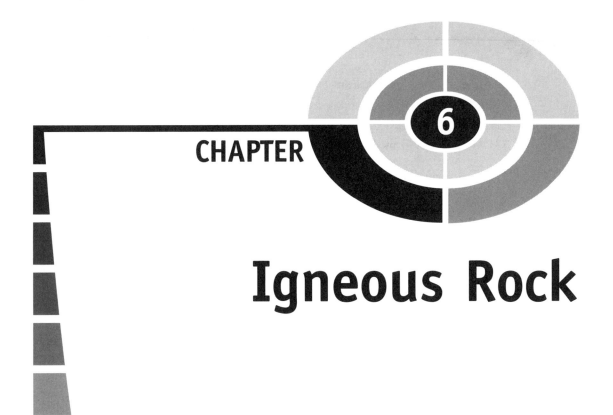

CHAPTER

6

Igneous Rock

Unlike plants and animals, most rocks have long histories. They seem ancient and never changing because within our lifetimes, they don't change much. A rock is a rock is a rock.

Igneous rocks, however, are probably the only rocks that give us a window into new rock formation. Igneous rocks are formed from magma that is sent through volcanic activity to the surface. Depending on the speed and way magma reaches the surface, the hardened igneous rock that is formed can look very different. We will look more closely at the three major magma types when we study volcanoes in Chapter 11.

There are three main rock types that come from magma: sedimentary, igneous, and metamorphic. Of these three, igneous is probably the most active and exciting. Igneous rock is created by exploding volcanoes and boiling undersea fissures. It has lots of distinct textures and colors depending on its chemical content and formation.

Six minerals make up nearly all of igneous rock. These minerals are quartz, feldspar, pyroxene, olivine, amphibole, and mica. The chemical elements that make up these minerals are silicon (Si), calcium (Ca), sodium (Na), potassium (K), magnesium (Mg), iron (Fe), aluminum (Al), hydrogen (H), and oxygen (O).

> Rocks formed by the hardening of molten rock (magma), whether deep in the Earth or blasted out during an eruption, are called **igneous rock**.

Igneous rock is formed from the cooling and hardening of magma within the Earth's crust. Over 95% of the top 10 miles of the crust is made up of igneous rock formed from lava eruptions. The root word *ignis* in Latin means fire. It is formed in temperatures of at least 700°C, the temperature needed to melt rock. The deepest magma in the mantle, next to the super-heated extreme heat of the outer core, has a different makeup from magma just beneath the crust and squeezed up through cracks or conduits.

The study of igneous rock is a study of *magma*, since igneous rock comes from cooled magma that has made its way to the Earth's surface. But not all magma is created equally. Depending on the time of heating and method of getting to the surface, different cooled magmas form rocks that look very different from each other. When scientists started studying igneous rock in the laboratory, they found two simple ways to separate igneous rock samples, *texture* and *composition*.

Sitting around a campfire on a starry night, the encircling rocks around the fire don't usually melt. It takes very high temperatures to melt rock. The type of rock-melting heat that affects igneous rock is a lot like that found on the Earth in its earliest days. From earlier chapters, we learned that the deeper into the Earth you go, the hotter the temperature. Sample temperatures taken at different depths commonly increase about 30°C per kilometer (90°F per mile). Of course, rock samples taken near magma lakes, along known fissures and near volcanoes, are a whole lot hotter.

> The rate of temperature increase compared to depth is known as a **geothermal gradient**.

Pressure also has an effect on the melting of rock. The greater the pressure applied to a solid (rock), the more force is applied to its atoms. This force packs the rock into denser and denser structures. Rocks deep in the mantle are under a lot of pressure. When a tectonic plate shifts or a crustal fissure forms releasing some of the overlying pressure, tightly compressed rock structure loosens up. Atoms aligned and held in a certain pattern within the rock structure are then able to shift. Their movement becomes freer and a lot more like a liquid state.

For example, the compound albite melts at 1104°C at the Earth's surface where the pressure is 1 bar. The melting temperature at 100 km, where the pressure is 35,000 times greater, is 1440°C. The extreme heat that couldn't affect the deeply pressurized rock can melt the less-compressed rock at the surface, allowing it to flow as a fluid.

Magma

What is magma, anyway? From Chapter 3 we learned that magma is the sea of melted rock found in the mantle. This super-heated liquid is hotter and cooler depending on its location and activity within the mantle's circulation currents.

Geologists use *pyrometers* to measure the temperature of lava from a distance. A *pyrometer* is an optical measuring device that allows temperature measurements to be taken safely. Freshly blasted magma has been measured at temperatures between 1000 and 1200°C. Once magma arrives, it cools. The cooler lava gets, the greater its viscosity and the slower it moves. But don't get too close, lavas that are barely moving have been measured at temperatures of 800°C.

> **Viscosity** is the resistance that a fluid has to flow because of its chemical and structural composition.

Temperature plays a big part in magma's viscosity. Think of pancake syrup or molasses; the hotter it gets, the runnier it gets. Heat excites the atoms and adds energy.

With magma, the silica content is also a big factor. Silicate minerals have a basic tetrahedral (pyramid-shaped) structure. They are linked together by shared oxygen molecules. However, silicate molecules in hot magma form crazy chains, sheets, and big matrices. As these linked silicate molecules get larger, the magma becomes more and more viscous and doesn't want to flow. The number of tetrahedral bonds that can be formed into linked molecule groups depends on the amount of silica present in the magma. Put simply, silica in magma gets hard when it cools.

> ↑ the number of tetrahedral, ↑ the linked group silicates,
> ↑ the viscosity

Some temperatures recorded at different sample locations are hotter in some areas of the crust than others. This tells geologists that the thickness of the crust changes and produces more volcanic activity in some areas than others. In active volcanic areas like Hawaii, the temperatures at 40 km have been recorded as high as 1000°C, while in more stable areas, the temperature at the same depth is only 500°C. After magma flows from the depths of the mantle out onto the crust, it is called *lava*.

Magma chambers are pockets of molten rock formed in the lithosphere. These chambers may be formed as surrounding rock is pushed down during plate interaction and melted. The outline of magma chambers have been seen while recording earthquake waves from active volcanoes. The depth, size, and overall shape of magma chambers can be figured out based on these readings.

> **Magma** is the origin of all volcanic rock. It has been around since the formation of the Earth.

When scientists studied the texture of quickly cooled magma, they found it took on two distinct forms: fine crystalline rock or glassy rock with no visible crystals. This is the magma blown violently from volcanoes during eruptions.

Some magma is very fluid and rains down fine molten fire, which cools quickly into ash. Some magma mixes with groundwater and creates super-heated steam and land-leveling mudflows.

Lava that slowly blobs out in bubbles and globs like slow moving molasses has a different texture. Since slow flowing lava streams and lakes below ground have a longer trip to the surface, they have time to form crystals. The longer lava cools, the larger and more complex the crystals can grow without interruption.

Rock Texture

In the late 1700s, while working in a field near his home in Scotland, James Hutton noticed coarse-grained granites cutting across and between layers of sedimentary rocks. Wondering how they penetrated the smooth fine sediments, Hutton thought they might have been forced into and between cracks as liquid magma.

The rock's texture provided Hutton with clues that the different rock types came from a different beginning.

As Hutton studied more and more about granites, he concentrated on sedimentary rocks that shared a border with the coarse granites, compared to sedimentary rock where no granite was present.

Hutton thought that the physical changes he saw in bordering sedimentary rocks must have come from an earlier exposure to high heat. This gave him the idea that molten magma from deep within the Earth had squeezed into areas of sedimentary rock and crystallized.

> **Grain size** and **color** are the two main ways that geologists describe rock textures.

The size of the minerals or crystals that make up a rock's texture is called *grain size*. Color can change depending on lighting, mineral content, and other factors, so it is thought to be less dependable when describing a specific rock.

When a rock's grains can be easily seen with the eye, roughly a few millimeters across, they are classified as *coarse grain*. When individual grains are not visible, the texture is considered to be *fine*.

Mineral grains or crystals have an assortment of different shapes and textures. They may be flat, parallel, needle-like, or equal in every direction like spheres or cubes. The shapes that crystals take, along with their grain size, combines to make rock samples unique. Think of it in terms of people and cultures of the world. Just as the combination of genetic inheritance and environment makes people individual and unique, the same thing happens with rocks!

Granites have a coarse grain size compared to obsidian with a very fine grain size. Granites are used for building materials because of their larger grain size and decorative pink or gray color. Obsidians are used for jewelry and art.

Intrusive Igneous Rocks

The cooled, crystallized magma that forces its way into the surrounding unmelted rock masses deep in the Earth is called *intrusive igneous rock*. It can be identified by its interlocking large crystals (grains 1 mm or larger), which grow slowly as magma cools over time. These are the large crystals Hutton observed. Commonly, large mineral grain igneous rock is known as *phanerites*.

When this type of rock is formed several kilometers below the surface of the Earth, it is known as *plutonic rock*. All masses of intrusive igneous rock, whether large or small, are called *plutons*. They are created on the slow boat of movement to the surface and have a lot of time to solidify and develop their individual unique compositions.

Plutons are given specific names depending on their size and shape. *Dikes* are areas of intrusive igneous rock that thrust up through other rock layers. They are generally perpendicular to the layers above, found at any depth, and trace the last push of a finger of upward rising magma. A *volcanic neck* is different from a dike in that it is discordant. It forms the feeder pipe just below a volcanic vent. The 400 m upthrusting mass of igneous rock known as Shiprock, New Mexico, United States is an ancient volcanic neck.

Sills are areas of intrusive igneous rock that are parallel to the layering of other intruded rock layers. They form flat, horizontal pockets between piles of rock layers at shallow depths and lower overlaying pressure. A *laccolith* is a mass of intrusive igneous rock that has pushed up between rock layers and been stopped to form a dome-shaped mound that looks like a blister.

Dikes and **sills** are often found together as part of a larger pluton network of intrusive igneous rock called a **batholith**.

Large plutons with outcrop exposures (rock sticking up through the ground) of greater than 100 km^2 are called *batholiths* and are huge compared to dikes and sills. They cover thousands of square kilometers and stretch across big parts of states and even between countries. One such monstrous batholith is the Coast Range Batholith that stretches from southern Alaska down the western coast of British Columbia, Canada to end in the state of Washington. It is roughly 1500 km in length.

When a batholith has outcrop exposure of less than 100 km^2 in length, it is called a *stock*. A stock of igneous rock is often found as a minor collection of rock located near the main batholith or as part of a mostly worn-away batholith.

Batholiths are huge masses of intrusive igneous rock, usually granite, with an exposed surface of larger than 100 km^2 and formed in the subduction zone along continental plate borders.

Granite, an example of intrusive igneous rock that crystallized slowly from magma below the Earth's surface, makes up a large portion of plutons and batholiths. Geologists' measurements of large batholiths have recorded depths of between 15 and 30 km thick. Figure 6-1 shows the differences between the creation of igneous intrusive and extrusive rock.

A single magma can crystallize into an assortment of igneous rock types. It doesn't solidify into one compound like water does when it freezes into ice. When magma solidifies, it forms many different minerals, which all crystallize at different temperatures. The different crystals solidify from the liquid magma, when their crystallization temperature is reached. Like a row of dominoes, as the temperature drops, crystals form one after another. Cooling magma then contains some fluid rock and some rock that has already hardened into crystals. When this happens, the concentration of certain minerals in the remaining magma is increased.

Sometimes when crystals are forming from an isolated chamber of magma, they are denser than the surrounding fluid. When this happens, they sink to the bottom of the chamber and form a separate crystalline layer with characteristics different from the remaining magma. As cooling, crystallization, and sinking of minerals continues, then many crystal layers with different compositions are formed.

In the early 1900s, geologist Norman Bowen was the first person to understand the importance of the temperature and the formation of separate crystals from magma. His studies at the Geophysical Laboratory in

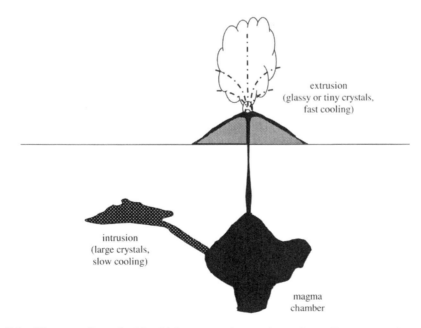

extrusion
(glassy or tiny crystals,
fast cooling)

intrusion
(large crystals,
slow cooling)

magma
chamber

Fig. 6-1.   The rate of speed with which magma rises to the surface affects crystal formation.

Washington, DC, which focused on the melting and crystallization properties of minerals, showed that as magma cooled at different temperatures, the composition of later formed crystals was very different. Bowen found that early forming crystalline rock had a lot more calcium, than later formed crystalline rock. As time went on, other geologists got interested in Bowen's ideas and the process of separating crystalline fall out from liquid magma became known as *magmatic differentiation by fractional crystallization.*

Extrusive Igneous Rock

Once magma exits the crust (ejected during an eruption) and cools rapidly, it creates a finely textured or glassy rock with small crystals. Basalt is an example of igneous rock that is quickly cooled from magma and extruded at the surface from lava. *Extrusive igneous rock* is on the rocket ship of movement to the surface. These rocks are commonly called *volcanic rock*, since they blast to the surface as either lava or rock fragments from your local neighborhood volcano. When the grains in igneous rock are not easily seen, even with a magnifying glass, the rocks are called *aphanites.*

> **Volcanic glass** is a natural glass formed by the quick cooling of molten lava that hasn't had time for crystals to form.

Depending on the cooling rate and amount of different gases trapped in cooling lava, volcanic glass can be smooth or full of holes. Most volcanic glass is in one of the three forms:

(a) *pumice*, a light-weight rock with lots of holes from escaped gases,
(b) *obsidian*, a glassy-smooth, dense solid, or
(c) *porphyry*, a mixed texture rock with large crystals suspended in a fine crystalline matrix. (This type is neither aphanite nor phanerite, but uniquely textured.)

The different textures of the three volcanic glass types are illustrated in Fig. 6-2.

Pumice rock is a favorite among movie set designers and practical jokers. It looks pretty much like regular rock, but it is very light. People can pretend superhuman strength when lifting a large pumice boulder. Pumice is sometimes called a glassy froth, like a molten milk shake. Since pumice is

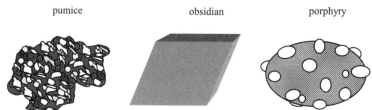

pumice   obsidian   porphyry

Fig. 6-2. Volcanic glass can be very different in appearance.

full of closed air pockets, it's really light. Pumice rock can even float! Impossible, right? No. There is enough trapped air to keep the pumice afloat.

When lava blasts out of a volcanic vent or fissure with terrific force and heat, the surrounding air is trapped with the exploding volcanic particles. After quick cooling, a lot of air is sealed within holes in the rock. Some pumice contains almost more air than it does rock. It has a rough uneven texture. Commonly, pumice stones are used as an abrasive tool in the beauty industry to smooth rough heals and calluses.

Obsidian is a black or dark-colored glassy volcanic rock, much like granite in chemical makeup, but formed by super-fast cooling when shot to the surface at low pressure during an eruption. It is the most "glassy" in appearance of the three types of volcanic glass types. It is shiny and smooth to the touch, but sharp at the edges.

Since it fractures fairly easily with light pressure along curved planes within the rock, obsidian was a favorite stone for early flint knappers. Its razor-sharp fracture edges made obsidian perfect for early knife blades, scrapers, spear, and arrowheads. In abandoned Native American settlements in the western United States and elsewhere in the world where early peoples lived, obsidian can be found lying on the ground in chips and fragments.

Obsidian is also excellent for dating ancient artifacts such as tools. By using a technique called *obsidian hydration-rim dating*, scientists can date tools from periods like the Aztec age in Mexico or preceramic Japanese era dating from 23,000 BC. The way it works is by testing for the presence of a *perlite* rim, formed when water molecules on the outside of the sample move inward (hydrating the sample) through cut obsidian edges. When this happens, the obsidian at the edges change to perlite. Most perlites have more water molecules than obsidian. The thickness of the perlite rim allows scientists to figure out how obsidian was shaped by human hands long ago.

Porphyry rock has some specific minerals most commonly associated with its igneous origin, but in general, is thought of as a smooth igneous rock with large crystals thrown into the rock like marshmallows in hot cocoa. These large crystals, called *phenocrysts*, formed while magma is still below the

Earth's surface. Like phanerites, they are shot to the surface during a volcanic eruption. Phenocrysts have well-formed crystals since they were created within fluid magma and didn't compete with other crystals growing and crowding them into warped shapes with irregular grain boundaries. We will look at these volcanic glass types in more detail in Chapter 11.

Basalt, a fine-grained, aphanitic, extrusive igneous rock is found everywhere under the sediments of the oceans' floors. It's like the wood or concrete floor of a house that lies under the carpet. The main minerals found in basalt are olivines, pyroxenes, and plagioclase feldspar. Basalt is the most abundant extrusive igneous rock on Earth.

Chemical/Mineral Composition

Igneous rocks are divided into two types depending on composition: *felsic* and *mafic*. Felsic rock is affected by heat, either from magma coming to the surface from extreme depths in the Earth or by the friction between continental plates. Although igneous rock contains some combined rock like the deep continental plate rock melted by moving magma, overall igneous rocks are either felsic or mafic.

Felsic

The first igneous rock type, *felsic*, is made up of light-colored igneous rocks that have high levels of silica-containing minerals like *quartz* and *feldspar*. Plagioclase feldspar that is higher in calcium crystallizes at higher temperatures than plagioclase having higher levels of sodium. When a rock is formed by different minerals, it tends to melt at a temperature below that of any one contributing mineral. This happens because different elements have different chemical properties that have an effect on their melting points.

Granite and *granodiorites* are the best known and most frequently seen intrusive igneous rocks containing about 70% silica. These mostly light-colored quartz and orthoclase feldspar minerals are found only in the continental crust. When different minerals like quartz and feldspar mix with granite, it takes on its well-known gray or pink color.

> **Felsic rock** has high levels of silica in its composition. The name *felsic* comes from a combination of the words feldspar and silica.

Rhyolite has the same composition as granite, but it is an extrusive igneous rock. It has a pale brown to gray color and is finely grained. The majority of rhyolites are made up of volcanic glass with no obvious crystals. They are much less common and found in much smaller pockets of extrusive igneous rock than their intrusive cousins.

Intermediate Igneous Rock

The volcanic igneous rocks in the intermediate class, that are a lot like granodiorite, are *dacite* and *andesite*.

Granodiorite has a lot of quartz like granite, but not a lot of silica. *Diorite*, the phaneritic cousin of andesite, contains feldspar and a lot of other mafic minerals like biotite and pyroxene that give it a darker color. Diorite is in between the granites and *gabbro* in composition and properties.

A transition rock type between the rhyolitic and basaltic magmas is an intermediate igneous rock like *andesite*. This type doesn't contain particularly high or low levels of silica, but is pretty average. Andesite is named after the volcanic Andes Mountains of South America and is made up of plagioclase and amphibole. Andesites and diorites are both equally common.

Mafic Rock

Mafic rock contains high levels of magnesium and ferric (iron-containing) minerals. The word *mafic* comes from a combination of these two mineral names.

Rocks which are low in silica, but high in magnesium and iron, form dark-colored mafic rocks like pyroxenes and olivines. Their main minerals are calcium-rich feldspar and calcium- and magnesium-rich pyroxenes.

The Earth's ocean floor is made up of basalt. Mafic rock contains only about 50% silica by weight. It is commonly dark gray, but can be green, brown, or black.

Figure 6-3 lists the main minerals found in felsic and mafic rock according to their composition.

When geologists looked closer at these two rock types, they found rocks that looked very different, but had the same composition. Some geologists thought that texture differences must be related to the way magma crystallized and then reached the surface, either by slow boat or by rocket ship.

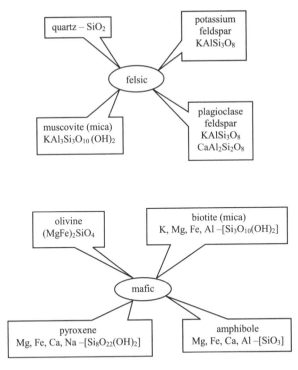

Fig. 6-3. Felsic and mafic rock have different amounts of silica and magnesium.

Crystallization

Since most magmas form in the upper mantle, their composition is mostly silica (SiO_2) with different amounts of iron, calcium, sodium, potassium, aluminum, magnesium, and other trace elements. The crystallization of these compounds is known, but as the ratios of the elements are different, the crystallization looks different.

In 1912, Bowen performed a series of experiments where he compared the crystallization temperature of compounds with different silica levels. He found that minerals, which crystallized at higher temperatures (calcium-rich *plagioclases*, *olivines*, *pyroxenes*), were low in silica. These high-temperature minerals were further divided into a ferromagnesian class and plagioclase feldspar. The minerals that crystallized at lower temperatures were usually low in silica. In 1928, Bowen published, *The Evolution of Igneous Rocks*, where he focused mostly on magma. Bowen became known as the *Father of Canadian Geology* for his ideas on crystal formation. This mineral

crystallization cycle is known as *Bowen's Reaction Series*. Figure 6-4 shows the minerals of the Bowen Reaction Series.

The plagioclase class is a *continuous reaction series* where some crystals are already formed, while the rest melt. The magma composition changes continuously, but crystals that are already formed don't change. Remember, the higher the amount of calcium in a crystal, the higher the temperature it takes to melt. As the magma cools, the crystals are constantly reacting with other elements in the melt. When cooling happens quickly, then the series shifts from high calcium-containing crystals to high silica-containing crystals.

The ferromagnesian class goes through a *discontinuous reaction series*. These elements begin with olivine crystallizing first, and then react with other elements in the magma melt to form pyroxene.

$$Mg_2SiO_4 + SiO_2 \rightarrow 2MgSiO_3$$

olivine + silica in fluid magma $\rightarrow$ pyroxene

As the magma cools and the temperature lowers even more, pyroxenes continue to react with elements in the melt. They are then converted to *amphiboles*. This series of reactions is interrupted (discontinuous) between each formation of different compounds. Some compounds are formed at

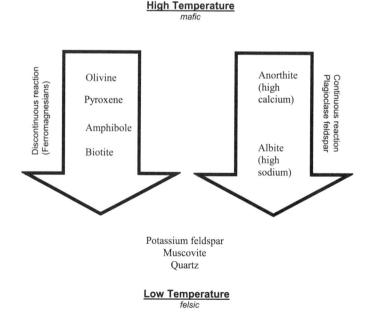

Fig. 6-4. The Bowen reaction series have continuous and discontinuous paths.

different temperatures, before they react with the melt elements and possibly form other new compounds with different compositions.

For example, at lower temperatures, pyroxene reacts with an increase in available silica and forms amphibole. The magma continues to cool and amphibole reacts to form *biotite*, which contains even greater amounts of silica.

Continuous and discontinuous reaction series are separate, but the external environment also has a role. Many times, the final form of the created rock can have unique characteristics that have been brought about by the environment specifics.

Some of the environmental factors that affect crystallization are: pressure, temperature changes, rate of cooling, local deposits of calcium- or sodium-rich minerals, and timing of crystallization during cooling. Magma can also be changed by the type of rock pockets that may be around it.

Magnetism

Certain minerals, the most important of which is *magnetite*, can become permanently magnetized. This comes about because the orbiting electrons around a nucleus form an electric current and create a very small magnetic field. Figure 6-5 shows how orbiting electrons set up an electric field. A *magnetic field* is the space through which the force or influence of a magnet is applied.

Above a temperature called the *Curie point* or *temperature*, the thermal excitement of atoms does not allow them to become permanently magnetized. They are too busy zipping around to slow down and allow magnetism to take place.

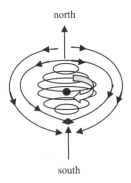

Fig. 6-5. Electrons orbiting a nucleus create a magnetic field.

> The temperature above which all permanent magnetism is destroyed is called the **Curie point** or **temperature**.

Curie's law, named after Pierre Curie, who with his wife, Marie, received Nobel prizes in chemistry for their work with radioactive elements. Curie's law describes the ability of an element to be magnetized as inversely proportional to the absolute temperature. In other words, the hotter it gets, the more the atoms get excited and the less likely magnetism is to occur.

The Curie point for magnetite is 500°C. Any temperatures higher than that cause atoms to get excited and vibrate wildly in no particular direction. This random dancing around causes the atoms' electrical currents to cancel each other out instead of lining up and forming a stable field.

When the temperature is less than 500°C, small "islands" of electric current in a solid stabilize and reinforce each other. When an external magnetic field is nearby, all the magnetic islands in a solid, parallel to the magnetic field, become larger and expand, taking over the neighboring, non-parallel islands. In no time, parallel islands of current form a "continent" of electric current and a permanent magnet is created.

This is true for cooling lava. All the minerals crystallize at temperatures above 700°C, a lot higher than the Curie points of any of the magnetic lava minerals. As crystallized lava slowly cools, its temperature drops below 500°C, the Curie point for magnetite. When this happens, all the magnetite grains in the rock turn into tiny permanent magnets. They are affected by the much greater magnetic field of the Earth.

Geologists have discovered from core samples of ancient lava and modern-day lava flows that the magnetic poles of the magnetite grains in the lava sample have the same magnetic inclination as the Earth's magnetic field.

Figure 6-6 shows how the magnetic field "islands" look above and below 500°C.

The magnetic poles of the magnetite grains in the lava will align in the same direction as the Earth's magnetic field. When lava samples are collected, they have unique magnetic polarities depending on the time and the magnetic field of the Earth that was in place when they were formed. The magnetic signature of a lava's formation will stay the same as long as the lava exists. The signature will be the same as when the lava's cooling temperature passed 500°C.

Figure 6-7 shows the difference that an external magnetic field makes when the magnetite grains are below the 500°C Curie point.

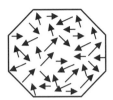

temperature higher
than 500°C

temperature lower
than 500°C
(*no external
magnetic field*)

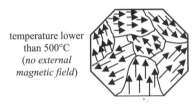

Fig. 6-6.   Atoms in tiny "islands" cause the formation of more magnetic islands.

temperature
below 500°C
(*external
magnetic field*)

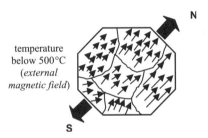

Fig. 6-7.   In an external magnetic field, most of the magnetic "islands" align with the external field.

This study of the magnetism of lava crystals allowed geologists to understand the changes that took place during the development and cooling of the Earth.

> When the cooling of lava results in the creation of permanent magnetism, it is called **thermoremanent magnetism**.

No magnet is permanently magnetic. Over time, it loses magnetism. This is called the *magnetic relaxation time*. Permanent magnets have very long relaxation times.

Relaxation of magnets is affected by many things including the following:

- mineral composition,
- grain size,
- temperature,

- neighboring minerals, and
- the strength of the original magnetization.

> The time it takes for a magnet to lose its magnetic ability is called the **relaxation time**.

In order for geologists to determine relaxation times and magnetization ages of rock samples, a few of the magnetic grains (perhaps those that were somewhat weaker to start with), must have already relaxed beyond the age of the first magnetization. Measuring the relaxation times of rock samples in the laboratory is performed as a temperature function.

Many igneous rock samples have relaxation times much greater than the magnetization age. These samples, collected from ancient, exposed lava flows around the world, are used to figure out where the magnetic poles were located thousands of years ago.

MAGNETIC REVERSAL

In 1930, Motonori Matsuyama, a Japanese mathematician and physicist, began studying magnetism in rocks. He took a closer look at the reason some rocks pointed in one direction, while others pointed in another. Matsuyama studied magnetic anomalies and proposed the idea that they were the result of *magnetic reversal*.

When geologists took samples of lava flows in Hawaii and other places, they found that some lava samples contained grains with reversed polarity. This meant that thousands of years ago the northern magnetic pole was located where the southern magnetic pole is today and vice versa.

> **Polarity reversal** is when the North Pole location and the South Pole location switch places.

The dating of lavas is possible through the use of radiometric methods using $^{40}K/^{40}Ar$ measurements. By using both radiometric dating and magnetic polarity measurements on ancient extruded lava layers, geologists were able to record the average between magnetic reversals. They found that, on average, the magnetic poles flip approximately every 200,000 years. By geological time the flip was overnight, but they actually happened over a gradual period of between 300 and 1000 years.

A magnetic switch begins with the reversal of an area of magma flow deep in the core of the Earth. As the switch area grows larger and more polar, the countercurrent works its way upward and begins to affect the magnetic currents in the crust and atmosphere. When this happens, areas of the Earth's outer magnetic field begin to weaken. The countercurrent below cancels out the charges above.

> **Magnetic field strength** or **field intensity** is the force applied to a magnetic pole at any point.

Weakened patches in a magnetic field are called *anomalies*. A *magnetic anomaly* may be high or low, subcircular, ridge-like, valley-like, or oval when studying a magnetic topographical map. The range of values of magnetic intensity over an anomaly or an area is called the *magnetic relief*.

The *South Atlantic Anomaly* is one of these weakened patches. In this area, the magnetic field is 30% weaker than other areas around the planet and it is growing. Geologists studying magnetic reversal over the past 10 years have used supercomputer programs, along with thousands of lava samples and the compass readings from British Naval officers' journal notes from the past 300 years, to study magnetism. The result is an excellent prediction method of magnetic reversal.

These studies have revealed that the Earth is long overdue for a magnetic reversal. The last major reversal happened over 700,000 years ago. Knowing this, geologists now think that the South Atlantic Anomaly is the beginning of a magnetic switch. It will not happen in our lifetime, but probably sometime in the next 1000 years if the model holds true.

Magnetic polarity can be minor or major. The tectonic and environmental effects of a magnetic reversal are not known. Scientists are just starting to study and understand the implications of a planet-wide magnetic reversal.

Times of mostly normal polarity, like what we have today, or times of mostly reversed polarity, are called *magnetic epochs* or *chrons*. The *Matsuyama Epoch*, a major polar reversal around 0.5–2.5 million years ago, is named after Motonori Matsuyama.

As lavas from many magnetic epochs pile on top of each other, they build up layers with opposite magnetic polarities. Figure 6-8 shows how these reversed lava layers might look if you were to take a cross-sectional slice.

Igneous rock provides geologists with many clues to the wild and crazy actions of ancient and recent magmas as they blasted or slowly forced their way to the Earth's surface in different magnetic fields. Studying

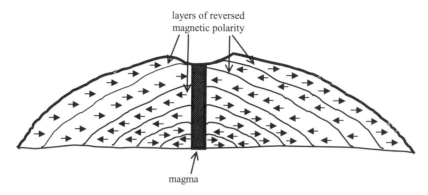

layers of reversed
magnetic polarity

magma

Fig. 6-8. Different lava layers contain igneous rock magnetized in reversed magnetic fields.

these clues will help us better understand magma's tricks and the Earth's future.

Quiz

1. The deeper into the Earth you go, the
   (a) dirtier you get
   (b) smaller the distance to China
   (c) hotter the temperature
   (d) colder the temperature

2. Granite contains approximately how much silica?
   (a) 30%
   (b) 50%
   (c) 70%
   (d) 90%

3. The majority of rhyolites are made up of volcanic
   (a) ash fragments
   (b) glass with no obvious crystals
   (c) sediments
   (d) coal particles

4. Rock that contains high levels of magnesium and ferric (iron-containing) minerals are called
   (a) mafic
   (b) sofic
   (c) felsic
   (d) ionic

5. The last major magnetic reversal happened over
   (a) 180,000 years ago
   (b) 320,000 years ago
   (c) 700,000 years ago
   (d) 900,000 years ago

6. What percentage of the top 10 miles of the Earth's crust is made up of igneous rock?
   (a) 52%
   (b) 75%
   (c) 95%
   (d) 99%

7. The cooled, crystallized magma that forces its way into surrounding rock is called
   (a) extrusive igneous rock
   (b) a bother, when it pops up in your corn field
   (c) ash
   (d) intrusive igneous rock

8. The *South Atlantic Anomaly* is
   (a) a math error in your vacation savings account
   (b) an island chain near Figi
   (c) a weakened area of the Earth's magnetic field
   (d) a strongly magnetic area of the Earth's magnetic field

9. Felsic rock has high levels of
   (a) silicon in its composition
   (b) felt in its composition
   (c) boron in its composition
   (d) tungsten in its composition

10. Extrusive igneous rock
    (a) slowly works its way to the Earth's surface
    (b) is often blasted to the Earth's surface
    (c) can be found in tar pits
    (d) always stays in the lithosphere

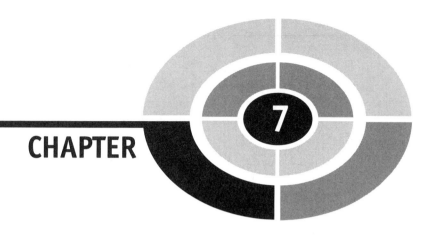

CHAPTER 7

Sedimentary Rock

At home, sediment is the leftover stuff that tends to build up, if it isn't cleaned up. It is the bits of string, the sand tracked in from the beach, the dog/cat hair and dandruff, and the playground gravel. It is the stray grape, the lost earring, the twig, and the broken pieces of cookie. It is made up of lots of stuff that were once whole or a part of something larger.

> **Sediment** is made up of loose particulate matter like clay, sand, gravel, and other bits and pieces of things.

I like to think of the formation of sedimentary rock as a lot like the gathering of dust bunnies in a house. Or how about the clothes, books, pens, papers, hair clips, shoes, CDs, and other stuff that collects on the floor of a teenager's room?

Sedimentary rock is formed when all the bits and pieces of different rocks, soils, and organic things are crunched together under pressure. They join into one tight mass that hardens into rock. It's a lot like when you leave oatmeal in a bowl too long. That stuff becomes concrete in a hurry!

> **Sedimentary rocks** are formed from rocks and soils that came from other locations and have become cemented together with the remains of dead organisms.

A collection of many different soil and rock types is known as *regolith*. Regolith is the loose rock material found scattered around the crust's solid, lower bedrock. It is made of volcanic ash, glacial drift, wind-driven deposits, plant accumulations, soils, and various eroded rock waste of every sort. Sedimentary rock that is carried by a glacier is usually deposited along underneath the ice or out to the sides.

Sedimentary rocks, originally from the buildup of material that gathers on the Earth's surface, have been compressed and cemented into solid rock over time. Most people can say they know something about sedimentary rock, since it is the most common rock type. Beautiful multicolored layers of sedimentary rock can be seen as rock walls when highways or railroads cut across hillsides and mountains exposing the different types of sediment layered there.

Sedimentary rock forms a broad blanket over the igneous and metamorphic rocks lying beneath its surface. In fact, Leonardo da Vinci wrote about the sedimentary rock of northern Italy in his journals. He compared sedimentary rock high on the mountainside with the sand and mud that he saw when visiting the Mediterranean coast. He noted that the color and textures were often the same.

Lithification

The word, *lithification*, comes from the Greek word *lithos*, meaning stone. Lithified soil is made up of sand, silt, and organic material. Lithification can take place soon after being deposited or much later. The rate of compaction and cementation also plays a big part in eventual lithification. Additionally, the heat needed for lithification is less intense than that found deeper in the mantle, so it's possible for lithification to happen in the top few kilometers of the crust.

> When sediment hardens into sedimentary rock, it is called **lithification**.

Grains squeezed together by the weight of overlying sediments during compaction are formed into rock denser than the original sediments. These dense layers are then sealed together by the precipitation of minerals in and among the layers.

This is how sandstone, formed from sand and limestone hardens after lithification with the hard skeletons and shells of marine organisms. When sandstone and limestone from different time periods are layered, the different texture and color of the layers can be easily seen.

> **Diagenesis** causes lithification of sediment by physical and chemical processes like compaction, cementation, recrystallization, and dolomitization.

Sediments become rock (*lithified*) through a combined process called *diagenesis*. Diagenesis is controlled a lot by temperature. But instead of the hot temperatures of igneous or metamorphic rock, diagenesis takes place at lower temperatures of around 200°C. Diagenesis can take place without complete hardening into rock, but you can't have lithification without diagenesis.

Diagenesis, which takes place in sedimentary rock after its first deposit, is specific to sedimentary rock during and after its slow hardening into rock. The word diagenesis is not used for weathering changes, soil formation, or metamorphism of rock into other rock types.

> The four main parts of **diagenesis** are: (1) compaction; (2) cementation; (3) recrystallization; (4) chemical changes (oxidation/reduction).

These usually take place in a stepwise way, but not always. They are often compressed and compacted first, then slowly cemented together by pressure that squeezes out water and air. Then, after being covered by more layers, unstable minerals recrystallize into a more stable matrix form or are chemically changed (like organic matter) into coal or hydrocarbons.

COMPACTION

Lithification through *compaction* is simple. As you pile more and more sediments on top of early laid-down sediments, the weight and pressure builds.

The heavier the weight, the more the lower layers get smashed together or compacted. The sediment's total volume is reduced since it is squeezed into a smaller space. Drying adds to the sediment's reduced volume.

Have you ever tried to guess the number of beans in a big jar? It isn't easy, but one thing you have to consider is how tightly the beans are packed within the jar. If there is a lot of air space between the beans, then the total number in the jar will be less. As more and more beans are added, they add weight at the top and pack the lower beans tighter. Some of the beans may even line up in the same direction. As even more pressure is added, the beans begin to crush and stick together. Eventually, with seasoning and some lemon juice, you get bean dip!

The same thing happens with compacted sediments. When shale grains are compacted and align in the same direction, they form rock that splits along a flat plane in the same direction as the flattened, parallel grains. Figure 7-1 shows how sedimentary particles are compacted when under pressure.

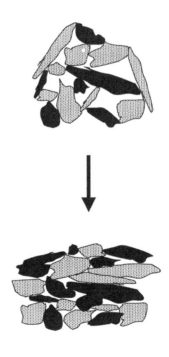

compaction

Fig. 7-1. Sediments pack together when weight and pressure are added.

CEMENTATION

Cementation of sediments happens when compacted grains stick together. Since most sediments are deposited in water, they have water molecules in the spaces between particles. The surrounding water contains different dissolved minerals that eventually fall out of solution and stick to the sediment grains. Minerals like calcite, silica, iron oxide, and magnesium cement the grains together into a solid mass that dries, is compressed further, and becomes rock.

Compaction and cementation can happen at the same time. The squashed sediments can be so tightly packed that they shut out the flow of mineral-containing water. Figure 7-2 shows how sedimentary particles are cemented and lithified as calcite and silica precipitate out of the surrounding water.

Additionally, minerals within the sediments can be dissolved away when water flows through. This creates pockets and places for other minerals or oils to gather. Petroleum geologists look for oil in these types of pockets.

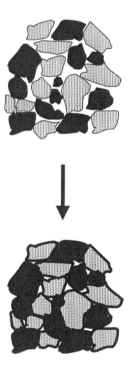

cementation

Fig. 7-2. Minerals in the water surrounding sediments can cement them together.

When sedimentary minerals dissolve and react with minerals in water to form other compounds, it is called *dolomitization*.

> **Dolomitization** happens when limestone turns into dolomite by a mineral substitution of magnesium carbonate for calcium carbonate.

CRYSTALLIZATION AND CHEMICAL CHANGES

Chemical and biochemical sediments and sedimentary rocks can be classified by their chemical makeup and properties. The ions of the most common elements dissolved into seawater are shown in Fig. 7-3. Although silica (SiO_2) and phosphorus play a big part in the makeup of sedimentary rock, they are only found in small amounts in seawater. When the water evaporates, the ions crystallize to form rock.

Carbonate sediments come from the biochemical precipitation of the decayed shells of microorganisms. Other chemical sediments that are high in calcium (Ca^{2+}) and bicarbonate (HCO_3^-) are precipitated out of seawater as calcium carbonate ($CaCO_3$) and carbonic acid (H_2CO_3) by inorganic processes and are much less common.

Types of Sedimentary Rocks

Unlike igneous rock, most sedimentary rocks have a fine-grained texture. Since a lot of the reason they have layered or settled in one place is due to water or wind, the particles of sediment are usually small and fine.

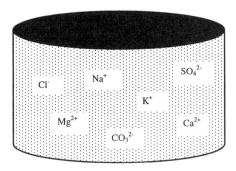

Fig. 7-3.   There are a variety of ions dissolved in seawater.

The way that sedimentary rock is deposited can also be related to size. Since wind can't blow or carry away boulders (well, maybe tornadoes can), generally it is the lighter, finer grains of silt that are transported by the wind.

In contrast to that, water tumbles rocks of different sizes. With the water deposit of sedimentary rock, current plays a big part. The stronger the current, the larger the rock and the farther it is carried. The relationship between current and particle size is the reason why many beds have the same types of particles. They sort and group according to size when flowing in the same current stream. So you see sand together with sand, river pebbles with other river pebbles.

CLASTIC

Clastic or *detrital* sedimentary rocks are formed from the weathering of existing rocks, which have been carried to a different spot from where they were originally and then turned into rock. They have a clastic (broken) texture made up of *clasts* (bigger pieces, like sand or gravel) and are grouped according to their grain size. Table 7-1 lists the various clastic particles and their associated sizes.

> **Detritus** is igneous, sedimentary, or metamorphic rock that has been moved away from its original location.

Table 7-1 Clastic sediments and rocks are named by size and shape.

Size of particle (mm)	Sediment	Rock
< 1/256th	Clay	Claystone or shale
1/256th–1/16th	Silt	Siltstone or shale
1/16th–2	Sand	Sandstone
2–64	Pebble	Conglomerate or breccia
64–256	Cobble or gravel	Conglomerate or breccia
> 256	Boulder	Conglomerate or breccia

Clastic sedimentary rocks are made up of pieces of other rocks. These pieces of rock are loosened by *weathering*, and then carried to some low area or crack where they are trapped as sediment. If the sediment gets buried deeply enough, it becomes compacted and cemented, forming sedimentary rock.

Clastic sedimentary rocks have particles ranging in size from microscopic clay to huge boulders. Their names are based on their clast or grain size. Beginning with the smallest grains, there are *clay*, then *silt*, then *sand*. Grains that are larger than 2 mm are called *pebbles*.

Shale is a rock made mostly of clay, *siltstone* is made up of silt-sized grains, *sandstone* is made of sand-sized clasts, and a *conglomerate* is made of pebbles surrounded by a covering of sand or mud. Figure 7-4 compares the different sedimentary rock types and their different proportions.

- *Coarse-grained clastics*
  Gravel (grain size greater than 2 mm; rounded clasts = conglomerate; angular clasts = breccia)
- *Medium-grained clastics*
  Sand (grain size from 1/16 to 2 mm)
  Sandstone (mostly quartz grains = quartz sandstone (also called quartz *arenite*); mostly feldspar grains = *arkose*; mostly sand-sized rock fragment grains = lithic sandstone (also called *litharenite* or *greywacke*))
- Cement (the glue that holds it all together) like calcite, iron oxide, silica
- *Fine-grained clastics*
  Silt and siltstone (grain size from 1/16 to 1/256 mm)
  Mud (clay), mudstone (claystone), and shale mud (grain size < 1/256 mm)
  Ironstone

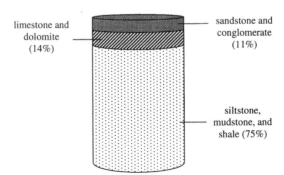

Fig. 7-4. Siltstone, mudstone, and shale are found in much higher amounts than other sedimentary rock types.

The deposit of these sedimentary rock types by different currents is as you might guess. The larger gravel, rocks, and pebbles are only carried along by strong currents. These are rushing mountain streams, rocky beaches with high waves, and glaciers' melt water. Strong glacial currents also carry sand. That is why you usually see sand between the gravel and pebbles. Pebbles and small stones are tumbled along and become smooth very quickly while bouncing along the land or in the water. Beach gravels and broken bits of glass, constantly rolled back and forth in the surf, also get smooth and rounded.

The coarse-grained clastic rock that doesn't easily smooth or erode is not a conglomerate, but instead a *breccia*. These sharp-edged rock fragments are found close to their source where sedimentary rock has been layered on top of them before they travel very far. Although some breccias are sedimentary in origin, others come from igneous rock and volcanic beginnings. They were deposited onto a sedimentary rock layer after first being shot out during an eruption or broken away from igneous rock along a fault during an earthquake.

SAND AND SANDSTONE

Sandstone is made up of mineral grains (mostly quartz) cemented together by silica, iron oxide, or calcium carbonate. Sandstones are commonly white, gray, brown, or red. Iron oxide impurities give the red and brown color to the darker colored sandstones. Most sandstones are gritty, while some are easily crushed (friable) and break apart to form sand. See Table 7-2 for the different types of sandstone.

The pores or spaces between the separate grains of sand in sandstones controls how porous the sandstone is. The amount and size of this spacing is called *porosity*. The porosity of sandstone allows sandstone to serve as good reservoirs for oil and natural gas. Petroleum engineers and geologists often

Table 7-2   Sandstone is classified according to the levels of its mineral content.

Majority of grains	Type of sandstone
Quartz	Quartz sandstone or quartz arenite
Feldspar	Arkose
Diverse rock fragments	Lithic sandstone (litharenite or greywacke)

look for these natural resources in sandstone areas. Sandstone is made up of one or more of the following:

- *Silt* (grain size 1/256 to 1/16 mm (gritty)),
- *Siltstone*,
- *Clay* (grain size less than 1/256 mm (smooth)),
- *Shale* (most abundant of sedimentary rock types),
- *Claystone*,
- *Mudstone* (a mixture of silt and clay or *mudshale* if it fractures along sedimentary lines), and
- *Ironstone* (clay ironstone).

Sandstones are very resistant to erosion and form bluffs, cliffs, ridges, rapids, arches, and waterfalls. Loose sands have many colors, but are commonly seen as white to light brown. Silica (quartz-rich) sands and sandstones of high purity (white color) are used widely in the glass industry for making window glass, light bulbs, vases, and utility containers. Tightly cemented sandstone is often used as a building stone.

Sand sediments are moved along by medium-speed currents like those of rivers, shoreline waves, and the wind. These can be rounded, which tells a geologist that they have traveled far (probably by water), or rough edged which usually points to shorter treks.

The amount of sand grains sorting in one area is another clue as to sediment origins. The average size of grains tells a lot about the strength of the current that carried them to a new spot and the type of parent crystal the grain was originally part of. If the grain sizes in sandstone are all the same, they are well sorted. If many grains are larger, with a lot of smaller and in-between grains, then they are poorly sorted. Sorting takes place during the sand's travels. Well-sorted sand grains commonly come from beaches, while poorly sorted grains are often the result of glacial travels.

ROCK ASPHALT

Rock *asphalt* is a medium- to coarse-grained sandstone with asphalt (*bitumen*) filling the pore spaces. It is squishy to solid, brown to black, has a pitchy to resinous luster, and is very sticky when completely saturated.

> **Bitumen** is made up of various mixtures of liquid, viscous, flammable, or solid naturally occurring hydrocarbons, excluding coal, that are soluble in carbon disulfide.

Rock asphalt deposits were formed when erosion of the surface rocks exposed oil-bearing rocks and allowed the more volatile hydrocarbons to escape. The asphalt-based crude gradually thickened until only a heavy tar or asphalt remained. The bitumen content of commercial rock asphalt varies from 3 to 15%. When asphalt is produced as part of some petroleum refining processes, it is called *artificial asphalt*.

Rock asphalt was once mined widely in Kentucky in the eastern United States. Large deposits were found in several areas. During the early 1980s, attempts were made to recover the petroleum in the rock asphalt by heat treatment, distillation, and other processes. Rock asphalt is used most commonly for surfacing streets and roads. It is also used for roofing, waterproofing, and mixing with rubber.

IRONSTONE

Iron oxide sediments are sedimentary rocks containing more than 15% iron in the form of iron oxides, iron silicates, and iron carbonates. Geologists think that most sedimentary iron was formed early in the Earth's history when there was less oxygen in the air and iron was more soluble. When soluble iron was carried to the ocean, it formed iron oxides and other compounds that then settled in layers to the bottom.

Ironstone, heavy, compact fine-grained stone is found mostly in nodules and in uneven beds with carboniferous and other rocks. It has 20–30% iron content and a clay-like texture, with large amounts of iron oxide, mostly *limonite*, in nodular form. Much of the iron produced in the United Kingdom is made from ironstone.

EVAPORITES

This group includes the *evaporites*, the *carbonates* (limestones and dolostone), and the *siliceous rocks*. Evaporites form from chemical elements dissolved in seawater. These compounds can be removed from saltwater and crystallized into rock by chemical processes or through biological processes (such as shell growth). Sometimes it's tough to sort between the two (carbonates and siliceous rocks), so evaporites are commonly grouped as chemical/biochemical.

> **Evaporites** that form as elements become more and more concentrated in an evaporating solution (usually seawater).

Marine evaporites are the sediments and sedimentary crystalline rocks formed as seawater becomes more and more salty through evaporation. Some marine evaporites are hundreds of meters thick. Huge amounts of seawater would have to evaporate for this amount of crystal formation to have been possible. Figure 7-5 illustrates how crystallization of evaporites happens in a shallow area with little freshwater input.

The most common types of evaporites include the following:

- *Carbonates* – mostly calcite and dolomite by diagenesis,
- *Gypsum* – made up of calcium sulfate ($CaSO_4$ and water),
- *Halite* (*rock salt*) – made up of sodium chloride ($NaCl$), and
- *Travertine* – made up of calcium carbonate ($CaCO_3$) (forms in caves and around hot springs).

Geologists have found that three things must happen in a bay for large amounts of evaporites to form. They are: (1) freshwater that flows into the bay from rivers and streams is limited, (2) connections to the open sea are constricted, and (3) the climate is parched and dry. In these bays, seawater evaporates constantly, but is replenished at a steady rate remaining supersaturated all the time. Evaporite minerals then settle steadily to the floor of the bay in sedimentary layers.

Phosphorite is another marine evaporite formed from chemical and biochemical sediments. It is made up mostly of calcium phosphate from places along the continental margins where ocean water is cold and deep. The *phosphorite* forms from an interaction between phosphate-rich seawater and muddy or carbonate-containing sediments.

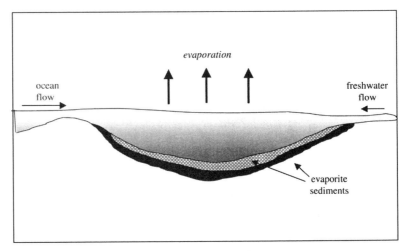

Fig. 7-5. Shallow basins allow the evaporation of crystals from super-saturated seawater.

Land (nonmarine) evaporites form in areas usually far from the sea. These are found in desert-region lakes with little or no river outlet. In these places, minerals come into the lake from chemical weathering and erosion, but without water current can't move on. One of the best known examples of this is the Great Salt Lake in Utah in the western United States. Rivers bring ions into the lake, but there they stay when the water evaporates. The concentrated dissolved ions in the Great Salt Lake make it one of the saltiest places on Earth, with levels eight times saltier than seawater.

Carbonate Sedimentary Rocks

Carbonate rocks all have carbon-related compounds in their composition. The two most important minerals found in carbonate rocks are:

- *Calcite* ($CaCO_3$)
- *Dolomite* ($CaMg(CO_3)_2$)

Carbonate sedimentary rocks are formed through chemical and biochemical processes. They include the *limestones*, which contain over 80% of the carbonates of calcium and magnesium, and *dolostones*. Limestone is made up of calcium carbonate ($CaCO_3$) from carbonate sands and mud, while dolostone is made up of calcium–magnesium carbonate ($CaMg(CO_3)_2$).

Dolomite formation is little different from some of the other evaporite and chemical sediments. Dolomite is formed by the reaction between sedimentary *calcite* or *aragonite* with magnesium ions in any seawater that trickles down through any sedimentary spaces. As the ions are exchanged, some of the calcium ions are switched with magnesium ions and calcium carbonate is then changed into dolomite.

Carbonate rocks are separated by their texture and contents. They include everything from fine mud to a mix-mash of fossils and debris. Table 7-3 lists the separate types and textures of carbonate sedimentary rock.

Unlike igneous rock, carbonate sedimentary rocks have a fine-grained texture. There are a variety of different forms found. Some of these include the following:

- *Micrite* (microcrystalline limestone) – very fine-grained; may be light gray or tan to nearly black in color; made of lime mud (*calcilutite*),
- *Oolitic limestone* (look for the sand-sized *oolites*),
- *Fossiliferous limestone* (fossils in a limestone matrix),
- *Coquina* (fossil hash cemented together; may resemble granola),

Table 7-3 Carbonate rocks are formed in a variety of ways.

Carbonate rock type	Composition
Micrite (microcrystalline limestone)	Very fine-grained; light gray or tan to nearly black
Oolitic limestone	Sand-sized oolites
Fossil laden limestone	Fossils in a limestone matrix
Coquina	Fossil mash cemented together; may resemble granola
Chalk	Microscopic planktonic organisms such as coccolithophores
Chert	Silica skeletons of sponges, diatoms, and radiolarians
Crystalline limestone	Larger grained than micrite
Travertine	Stalactites and stalagmites ($CaCO_3$)
Coal	Converted land plant remains
Other	Intraclastic limestone, pelleted limestone

- *Chalk* (made of microscopic planktonic organisms such as coccolitho-phores; fizzes readily in acid),
- *Crystalline limestone*,
- *Travertine* (evaporates of calcium carbonate, $CaCO_3$) stalactites and stalagmites, and
- *Other* – intraclastic limestone, pelleted limestone.

Siliceous Rocks

This type of sedimentary rock is commonly formed from silica-secreting organisms such as *diatoms*, *radiolarians*, or some types of sponges. It is most commonly called *diatomaceous earth*. Many expert gardeners use it to aerate and balance the acidity in soil with high silica containing diatomaceous earth.

> **Siliceous** (silica-containing) **rocks** are sedimentary rocks with high silica (SiO_2) content.

Biologic sedimentary rocks form when large numbers of living things die, pile up, and are compressed and cemented to form rock. Accumulated and pressurized carbon-rich plant material may form coal. Deposits made mostly of animal shells may form *limestone, coquina,* or *chert.*

Diatomite looks like chalk and fizzes easily in acid. It is made up of microscopic plankton (tiny plants) called *diatoms.* When the silica from diatom remains is dried and powdered, it is used as one of the main ingredients in *dynamite.*

Chert (also known as *flint*) is very different in appearance from diatomite. It is made of hard, extremely fine, microcrystalline quartz and can be dark or light in color. Chert is formed when silica in solution goes through chemical changes within limestone. It often replaces limestone and does not fizz in acid.

Flint was used by early hunters for spear and arrowheads. It was easily formed into points and sharp, cutting edges. *Opal* is a white or multicolored, less-developed crystalline form of chert used in jewelry. Opal has a high water content.

COALS (ORGANIC SEDIMENTARY ROCKS)

Organic sedimentary rock is made up of rocks that were originally organic material (like plants). Because of this, they don't contain inorganic elements and minerals. These organic sedimentary rocks are known as *coals.*

Coals are usually described in the order of their depth, temperature, and pressure. They are made almost completely of organic carbon from the diagenesis of swamp vegetation. Coals contain the following types of materials:

- *Peat* (spongy mass of brown plant bits a lot like peat moss),
- *Lignite* (easily broken and black),
- *Bituminous coal* (dull to shiny and black; sooty; sometimes with layers), and
- *Anthracite coal* (very shiny and black, a bit of a golden gleam; low density; not sooty; could be a metamorphic rock from exposure to high temperatures and pressures).

Coal is formed from peat, which is a collection of rotting plants found in and around swamps. The conversion from peat to coal is called *coalification.*

In the various stages of coalification, peat is changed to *lignite*, lignite is changed to *subbituminous coal*, subbituminous coal is changed to *bituminous coal*, and bituminous coal is changed to *anthracite coal*.

In the United States, coal is found in areas of eastern and western Kentucky, where it is layered between shales, sandstones, conglomerates, and thin limestones. The time span from approximately 320 million years ago and until about 30 million years is commonly called the Coal Age.

Sedimentary Stratification

We saw how sedimentary layers can gather in one location, as a result of natural processes such as waves, currents, drying, wind, and other factors in Chapter 5, when we looked at different stratas.

Geologists often use the words *sedimentary bed* and *sedimentary layer* to mean much the same thing. I have followed this pattern and will use both words to define sedimentary rock layers. The sedimentary rock strata are laid down in certain well-known structures such as:

- Lamination bedding,
- Uniform layers,
- Cross-bedding,
- Graded beds,
- Turbidity layers, and
- Mud cracks.

We will look at the differences between these types and how they give a different look to a variety of sedimentary rock layers.

BEDDING

Nearly all sedimentary rocks are laid down in layers or beds. Layers can be very thin, like a few millimeters or as thick as 10–20 m or more. This sedimentary layering or *bedding* gives it the characteristic striped look seen in the Grand Canyon and deserts of the United States. The exposed mesas and arches are made of layered sedimentary rock.

A *bedding plane* is a specific surface where sediments have been deposited. Bedding most often happens in a flat plane as wind or water has layered it over and over onto the same area. When a bedding plane has a different color from surrounding rock, it makes it easier to spot one layer from another. Although we think of bedding as horizontal, this is not always the case.

> **Bedding** is the formation of parallel sediment layers by the settling of particles in water or on land.

LAMINATION BEDDING

When an area has fine, thin (less than 1 cm in thickness) bedding layers, it is known as *lamination* or *lamination bedding*. Over millions of years, a single bed made up of very thin individual layers can be several meters thick. Different sedimentary lamination layers can be set apart by grain size and composition. These differences are caused by the different environments in place over long stretches of geologic time.

UNIFORM LAYER

A sedimentary rock layer, made up of particles, all about the same size, is known as a *uniform layer*. A uniform layer of clastic rock has particles of a single size that have been tumbled by a current of a constant speed. If a uniform bed is made up of layers of single particle sizes, it is thought that currents of different speeds caused the uniform layering of like particles at different times. When nonclastic minerals precipitate out of a solution, the crystals that form uniform layers are all the same size.

CROSS, GRADED, AND TURBIDITY BEDDING

Cross bedding happens when wind or water causes sedimentary layers to be laid down at inclined angles to each other. These can be up to 35° from the horizontal and are found when sediments are laid down on the downhill slopes of sand dunes on land or sandbars in rivers or shallow seas. Cross-bedding of wind-deposited sediments can be beautifully complex with many changes in direction. Figure 7-6 gives you an idea of cross-bedding found in sandstone.

Graded bedding comes about through the sorting of particles by a current. A graded or gradient bed is layered with heavy, coarse particles

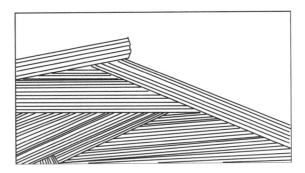

Fig. 7-6. Sandstone cross-bedding shows wind direction at the time sediments were deposited.

at the bottom, medium particles in the middle layers, and fine particles on top.

> A **graded layer** is made up of particles that are layered from coarse to fine with the heaviest particles on the bottom.

It's like a jar full of beach sand, small shells, sea glass, and seawater. If you shake it up, everything swirls around for a few minutes before settling. When settling according to weight, the heavier glass pieces settle first, followed by the shells, before everything is coated finally by sand. Over geologic time, these graded beds are piled on top of each other, many meters thick, by deep ocean currents along the sea floor. An example of a graded sedimentary rock bedding sequence is shown in Fig. 7-7.

Turbidity bedding is found as *ripples* in the sedimentary rock record. Just as parallel lines of beach sand near the water line are caused by the constant pounding of the surf, sedimentary rock layers are hardened in these same patterns.

When bedding of sediments happens in water, it is almost always horizontal. But currents can affect the look of sedimentary rock as well, with constant wave action giving sedimentary rock layers a symmetrical look. Water currents making swirls and eddies cause permanent overlaying of sedimentary rock. Waves at the beach, constantly depositing sediment with a back-and-forth movement, produce bedding with evenly shaped (symmetrical) peaks. Sediments deposited by a current in only one direction cause sedimentary peaks to be tilted away (asymmetrical) from the direction of the current. Figure 7-8 shows the differences between these two peak forms.

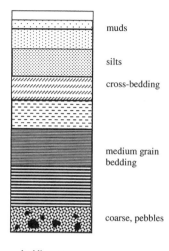

muds

silts

cross-bedding

medium grain
bedding

coarse, pebbles

bedding sequence

Fig. 7-7.   Bedding gradients reflect the size sorting of layered sediments.

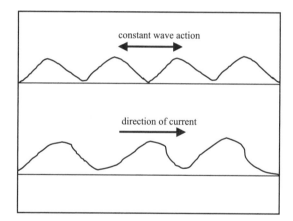

constant wave action

direction of current

Fig. 7-8.   Different sandstone ripple forms are caused by waves versus current.

BIOTUBATION

Sometimes you will see sedimentary rock with tubes crossing vertically or at an angle through several layers. This is known as *biotubation*. These fossilized sedimentary structures are the remains of burrows and tunnels made by worms, clams, and other marine organisms. These primitive ocean bottom-dwelling residents burrow through sedimentary layers in search of organic material. Geologists study their vacant homes and waste for clues to the ancient environment during the time they lived.

Sedimentary Environments

Sedimentary environments are places where sediments collect and sedimentary rocks form. They can be grouped into three main areas: terrestrial (land), marine, and transitional (border) environments.

1. *Terrestrial sedimentary environments* (land)
   (a) Rivers, streams, and ponds
   (b) Lakes
   (c) Swamps
   (d) Deserts
   (e) Glacial environment

2. *Transitional environments* (border areas between the land and marine environments)
   (a) Beach and barrier islands
   (b) Delta
   (c) Lagoons
   (d) Estuaries

3. *Marine environments*
   (a) Continental shelf
   (b) Continental slope and rise (deep sea fans)
   (c) Abyssal plain
   (d) Reefs

TERRESTRIAL SEDIMENTARY ENVIRONMENTS

The first of the sedimentary environments is the best known since most people have visited streams, rivers, or lakes at one time or another. A good amount of clastic fragments are deposited into sedimentary layers within *terrestrial sedimentary environments*. This happens through the action of water current or blowing wind. Depending on the way the sediments are laid down, different layering patterns are seen.

TRANSITIONAL ENVIRONMENTS

An in-between or transitional sedimentary environment is found where major sources of water currents meet the ocean. In *delta* areas, there is a rich mixture of sediments that arrive from all along the route of the current.

The mouth of the Mississippi River delta near New Orleans, Louisiana (United States) deposits many tons of silt into the Gulf of Mexico in a wide fan of sand and mud that can be seen from space.

Beach environments have a lot of wave and tidal energy that moves particles constantly. This back-and-forth grinding movement polishes and sorts them according to size. Fine sediments are washed away to settle further out in the tidal flats, where the wave action is less.

MARINE ENVIRONMENTS

Marine environments include fine sediments that settle to the ocean bottom as the remains of marine organisms and plants.

The finest marine sediments are found far from the continental margins. *Pelagic* (from the Greek word *pelagos*, meaning sea) sediments are so tiny as to be found suspended in salt water most of the time. Think of the super fine dust that is always settling out of the air onto furniture at home. You can't even see it unless a ray of sunlight makes it visible.

Pelagic sediments are made up of the calcium-containing shells of microorganisms such as *foraminiferans*, *radiolarians*, and *diatoms*. These microorganisms live near the ocean's surface and when they die, their shells sink, decay, and become part of the fine-grained mucky ooze on the ocean's bottom. Pelagic sediments, dispersed all through the oceans, settle out and form layers of fine sediment onto deep ocean plains.

Unique marine sediment is created by chemical precipitation in seawater. Precipitates of manganese oxides and hydroxides form golf ball to basketball size lumpy nodules strewn around the ocean floor.

When we talk about ocean environments in Chapter 13, we'll go into much greater detail on the currents, inhabitants, and characteristics of the planet's oceans. We will see what makes them different and the same in several areas around the globe.

Weathering

Water, wind, and ice all work together to breakdown solid rocks into small rocky particles and fragments. These bits of rock are swept away by rain into streams. Gradually these particles get deposited at the bottom of stream beds or in the ocean. As more and more sediment builds up, it gets crushed together and compacted into solid rock.

> **Weathering** wears away existing rocks and produces lots of small rock bits.

With every tick of the clock, day after day, rock surfaces are worn away by wind and rain. Small bits of dirt, sand, mud, and clay are slowly ground away and washed into streams, rivers, lakes, and oceans. After these tiny bits of sand and rock settle at the bottom, they become sediment.

Water minerals and microscopic or tiny organisms also get mixed with the dirt and sand to form sediment. Over time, more and more sediment piles up on top of what was there before. After millions of years the sediment builds up into deep layers. The heavy weight and extreme pressure from the constantly added sediment turns ocean sediment at the bottom into sedimentary rock. The oldest ocean sedimentary rocks are thought to be around 600 million years old.

These oldest sedimentary rocks were formed long ago, but since then, they have been crushed, heated, and transformed into what is known as metamorphic rock. We will see how metamorphic rock is different from igneous or sedimentary rock in Chapter 8.

Quiz

1. Lithification, comes from the Greek word *lithos*, meaning
    (a) light
    (b) white
    (c) stone
    (d) seawater

2. Deserts are an example of
    (a) marine environments
    (b) something eaten after the dinner meal
    (c) a variety of chert
    (d) terrestrial sedimentary environments

3. Transitional environments include all but which of the following?
    (a) beach and barrier islands
    (b) volcanoes
    (c) lagoons
    (d) estuaries

4. Grains measuring about 1/256 to 1/16 mm are called
    (a) breccia
    (b) sludge
    (c) silt
    (d) rice

5. The loose rock scattered around the solid, lower bedrock of the crust is called
    (a) dolomite
    (b) dirt
    (c) regolith
    (d) aethenosphere

6. Some sedimentary rock is formed from silica-secreting organisms called
    (a) diatoms
    (b) blue-green algae
    (c) flint
    (d) trilobites

7. Bedding is the
    (a) sheets and blankets on a bed
    (b) formation of parallel sediment layers by the settling of particles
    (c) soil used in landscaping
    (d) stuffing found in thick quilts

8. Diagenesis causes
    (a) a bright red rash on the elbow
    (b) miscalculations of circular core samples
    (c) lithification of sediment by physical and chemical processes
    (d) photosynthesis in most upper ocean layers

9. As elements become more and more concentrated in a solution they form
    (a) club
    (b) hydrocarbon chains
    (c) distillates
    (d) evaporites

10. Clastic sedimentary rocks are made up of
    (a) pieces of other rocks
    (b) gelatin from kelp
    (c) petroleum distillates
    (d) boulders larger than 1 m across

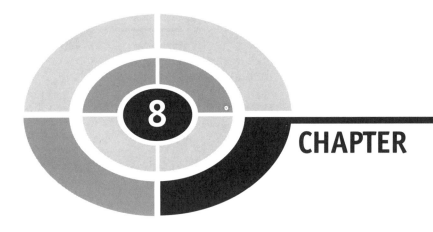

CHAPTER

Metamorphic Rock

Igneous rock is formed as a result of the Earth's internal "engine," while sedimentary rock formation depends on external climate and conditions. Metamorphic rock, however, takes place after these rock types have already formed. It is created by transforming igneous or sedimentary rock into something new.

Of the three major rock types, igneous, metamorphic, and sedimentary, *metamorphic rock* is the chameleon rock. It transforms into different types of rocks depending on the factors that it is exposed to within the Earth. This rock type is both a wonder and a headache to geologists. Since metamorphic rock begins originally as something else, it can be confusing as to whether it is the original rock or a transformed version. To solve this problem, geologists gather clues from the surrounding area or an outcrop from which the sample rock is found.

Besides being intruded by magma regularly, the Earth's crust is subjected to stresses within the crust and mantle that cause it to break and bend forming *fault folds*. These forces often center along thin, winding belts when folding. They also combine with magma intrusion and extrusion while pushing up mountain ranges. The rocks within a mountain range are not only under extreme pressure, but heated by magma intrusion as well. These stresses deform and recrystallize rock to different degrees. Pressure and temperature can also change previously metamorphosed rocks into new types.

Rock-forming and destroying processes have been active since the Earth was first formed. When sedimentary and igneous rocks are exposed to extreme pressure or medium heat, they are changed. They become *metamorphic rocks*, which form while deeply buried within the Earth's crust. It is important to remember that metamorphism does not just melt existing igneous or sedimentary rock, but transforms it into a denser, compacted rock.

> **Metamorphic rocks** are formed from rocks that were originally another type and were changed into a different form.

Metamorphic Rocks

The name *metamorphic* comes from the Greek words, *meta* and *morph*, which mean "to change form." Geologists have found that nearly any rock can become a metamorphic rock. When existing rock is shoved and pressurized, its minerals become unstable and out of equilibrium with the new conditions, causing them to change.

Remember the chameleon? When a chameleon moves from a gray rock to a bright green leaf, he changes his skin color to the same as his environment. By adjusting to his new conditions, the chameleon protects himself and comes into equilibrium with his surroundings. The process of metamorphism is similar. When a rock is slowly moved through tectonic processes to a new temperature or pressure environment, its original chemical and physical conditions are changed. In order to regain stability in the new conditions, chemical and physical changes take place. With metamorphism, mineral changes always move toward reestablishing equilibrium. Common metamorphic rocks include *slate*, *schist*, *gneiss*, and *marble* with many grades in between.

Most of the time, metamorphic rock is buried many kilometers below the crust which allows increasing temperatures and pressures to affect it. However, metamorphism can also happen at the surface. When geologists study soils under hot lava flows, they find metamorphic changes. The three main forces responsible for the transformation of different rock types to metamorphic rock are:

- internal heat from the Earth,
- weight of overlying rock, and
- horizontal pressures from rock that changed earlier.

Temperature

Temperature increases in sedimentary layers which are found deeper and deeper within the Earth. The deeper the layers are buried the more the temperature rises. The great weight of these layers also causes an increase in pressure, which raises the temperature even more.

This cycle of heat and pressure that describes the transformation of existing rock is called the *rock cycle*. It is a constantly changing feedback system of rock formation and melting that links sedimentary, igneous, and metamorphic rock. Figure 8-1 illustrates how the three rock types feed into a simple rock cycle. We will look at more of the factors that play into the rock cycle in Chapter 15 when we look at weathering and topography.

The pushing down of rock layers at subduction zones causes metamorphism in two ways: the shearing effect of tectonic plates sliding past each other causes the rocks to be deformed that are in contact with the descending rocks. Some of the descending rocks melt from this friction. These melted rocks are considered igneous rock not metamorphic. Then secondly, nearby solid rock that lies alongside melted igneous rock can be changed by high heat to also form metamorphic rock.

We learned in Chapter 3 that the temperature of the Earth increases the deeper you go. On average, the temperature increases 30°C/km, but can vary from 20 to 60°C/km in depth. For example, the temperature at a depth of 15 km is equal to 450°C. At the same depth, the pressure of the overlying rock is equal to 4000 times the pressure at the surface.

This heat and pressure gradient, changing with depth, allows metamorphism to happen in a graded way. The deeper you go, the hotter the temperature and pressure, the greater the metamorphic changes. Depending

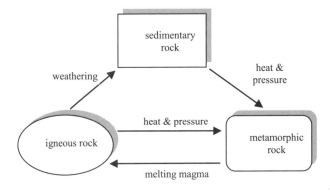

Fig. 8-1. The rock cycle is affected by environmental conditions of heat, pressure, and weathering.

on the conditions under which rock is changed, the rock gradient forms new metamorphic rock into high-grade or low-grade metamorphic rock.

↑ Temperature   and   ↑ Pressure ⇨ High-grade metamorphic rock
↓ Temperature   and   ↓ Pressure ⇨ Low-grade metamorphic rock

---

As rock adjusts to new temperature or pressure conditions, the crystal structure of its minerals are affected. Ions and atoms are energized. They begin breaking their chemical bonds and creating new mineral linkages and forms. Sometimes, crystals grow larger than they were in the original rock. New minerals are created either by rearrangement of ion bonding or by reactions with fluids that enter the rocks.

There are five main ways that metamorphic rocks are created. These different metamorphic rock processes include *contact*, *regional*, *dynamic metamorphism*, *cataclastic*, *hydrothermal*, and *burial metamorphism*. A closer look at each one of these will show how they are different.

Contact Metamorphism

Contact metamorphism takes place when igneous intrusion of magma heats up surrounding rock by its extreme temperatures. This surrounding rock is called *country rock*. When igneous intrusion happens, the country rock's temperature rises, heats up, and becomes permeated with fluid brought along by the traveling magma. The area affected by hot magma contact is usually between 1 and 10 km in size.

When contact metamorphism happens on the surface because of an outpouring of lava, it is restricted to a fairly thin rock layer. Since lava cools quickly and gives heat little time to penetrate the underlying country rock, the metamorphism that takes place is limited.

An *aureole* or rock halo is formed by metamorphosed rock around a high-temperature source. The metamorphic rock close to the magma pocket contains high-temperature minerals, while rock found further away has lower-temperature minerals. These heat sources are commonly closer to the surface crust in contact metamorphism than other types. Figure 8-2 shows the surrounding aureole effect of magma heating.

When a plutonic magma pocket is rimmed by a contact ring of metamorphic rock, it is known as an **aureole**.

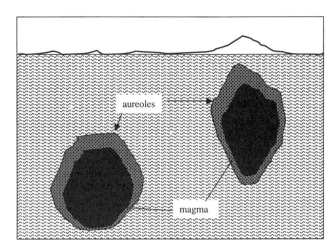

Fig. 8-2. Surrounding rock is heated in the area around magma pockets in a halo effect.

A special type of contact metamorphism, *impact metamorphism*, is caused by the high-speed impact of a meteorite. As the meteorite hits the Earth's surface, it causes shock waves. These are sent out from the impact site as a way to scatter the energy from impact. Depending on the speed and angle of impact, the surface at impact is immediately compacted, fractured, melted, and may be vaporized. Following the initial slam and shock wave, the rock decompresses sending rock flying in all directions and forming an impact crater.

Have you ever seen high-speed photography of a droplet of water hitting the surface of a still pool? The impact compresses the water's surface downward for an instant, followed immediately by a rebounding ring of droplets shooting upward. The shock-wave impact is absorbed throughout the liquid as ripples.

> Unlike deep mantle metamorphism, **shock metamorphism** happens in the instant of a high-velocity impact.

A meteorite impact has much greater velocity and energy than a free-falling droplet, but impacts in much the same way. For example, an iron meteorite measuring 10 m across and hitting the surface at a velocity of 10 km/sec would cause a crater over 300 km in diameter.

The shock wave from a meteorite impact causes high-pressure shock metamorphism effects such as specific fracture patterns and crystal structure destruction. In fact, the formation of *polymorphs*, or in-between shock-related minerals like *coesite* or *stishovite*, not commonly found on the surface, helps geologists to find ancient impact craters.

Contact metamorphism produces *nonfoliated rocks* (without any lines of cleavage) such as *marble*, *quartzite*, and *hornfels*.

> **Nonfoliated rock** is made up of crystals in the shape of cubes and spheres that grow equally in all directions.

Marble is formed from metamorphosed limestone or dolomite that has recrystallized into a different texture after contact with high heat. It is made up of calcite, but if it contains a large amount of dolomite, then it is called dolomitic marble. Both limestone and dolomite have large amounts of calcium carbonate ($CaCO_3$) and many different crystal sizes. The different minerals present during the formation of marble gives it many different colors. Some of marble's colors include white, red, pink, green, gray, black, speckled, and banded.

Since marble is much harder than its parent rock it can be polished. Marble is used as a building material, for kitchen and bathroom countertops, bathtubs, and as carving material for sculptors. Grave stones are made from marble and granite because they weather very slowly and carve well with sharp edges.

Quartzite is the product of metamorphosed sandstone containing mostly quartz. Since quartzite is formed from sandstone that contacted hot, deeply buried magma, it is much harder than its parent rock. As it is transformed, the quartz grains recrystallize into a denser, tightly packed texture. Unlike matte-finished sandstone, quartzite has more of a shiny, glittery look. While sandstone shatters into many individual grains of sand, quartzite fractures across the grains.

Hornfels is a fine-grained, nonfoliated, large crystal metamorphic rock formed at intermediate temperatures by contact metamorphism. These can be further defined as pyroxene-hornfels and hornblende-hornfels formed at still lower temperatures.

Figure 8-3 shows how magma forces its way into layers of limestone, quartz sandstone, and shale. The high heat coming from the deep magma chamber changes these sedimentary rocks into the metamorphic rocks, such as marble, quartzite, and hornfels. These changed rocks are listed to the right of the figure in relation to their original rock types.

Hydrothermal Metamorphism

This type of metamorphism is common with mid-ocean ridges where the crust is spreading and growing as a result of the outpouring of hot lava.

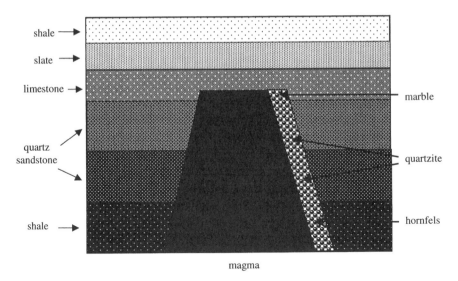

Fig. 8-3. Contact metamorphism happens when magma surges upward through existing rock.

The ocean water that bubbles through the hot, fractured basalts of the ridge margins becomes heated, causing chemical reactions between the surrounding ridge rock and seawater. These chemical changes produce metamorphosed basalt.

Hydrothermal metamorphism can also take place on land, when fluids from igneous rock intrusions percolate through surrounding country rock, causing a *regional metamorphism*.

Migmatites

Higher temperature and pressure metamorphic boundaries mark the lower limits of magma production. With a good amount of water, magma formation starts at a lower temperature. When there is little water, magma doesn't form until higher temperatures are reached. This allows different types of metamorphic rock (schists, gneisses, and amphibolites) to form in different areas depending on the amount of fluid present.

Different types of layering are also possible depending on fluid intrusion, as well as temperature and pressure factors. When there is a variety of metamorphic rock types in an area, geologists find that a combination (mixed) rock has formed. Alternating layers of granite and schist form a mixed rock called *migmatite*.

> A combination metamorphic rock type that contains both igneous and metamorphic rock is known as **migmatite**.

Burial Metamorphism

When layers of sedimentary rock become heavier and heavier, they get pushed further down into the crust, where they heat up and take on the temperature of the surrounding rock. We learned in Chapter 7 that when this happens, digenesis causes the transformation of sedimentary rock minerals and their textures. It happens at temperatures below 200°C.

As a result of increasing temperature and pressure in sedimentary rock layers, by ever heavier upper layers, diagenesis slowly continues and changes sedimentary rock layers over time through the process of low-grade *burial metamorphism*.

This type of metamorphism often causes partial mineral changes in sedimentary rock with some bedding layers left unaffected. Burial metamorphism usually causes wide folding of sedimentary rock layers within the greater changes of regional metamorphism.

Cataclastic Metamorphism

Cataclastic metamorphism takes place in the same areas as igneous activity along plate margins, oceanic, and continental hot spots, and deformed mountain ranges.

Tectonic plate movement causes high-pressure metamorphism by crushing and shearing rock away as a result of plate movement. When metamorphism happens along a fault, the transforming heat comes from intense friction and pressure going on between massive plates as they grind past each other.

Broken and metamorphic rock fragments found along a metamorphic rock fault are called *fault breccia*. This rock type has minerals that crystallize at either extreme temperature or the high pressure and low temperature associated with extreme frictional stress. This type of metamorphism is often part of regional metamorphism.

Regional Metamorphism

Regional metamorphism is the most widespread kind of metamorphism. This takes place over a much greater crustal area where both temperatures and pressures are high. Geologists use the term regional metamorphism when talking about large-scale metamorphism rather than that found locally near specific igneous rock intrusions or faults. Most regional metamorphism takes place in the deeper levels of the crust, along the margins of clashing and subducting tectonic plates, where rock is deformed and forced into a new direction. Regional metamorphism is fueled by the Earth's internal heat.

Regional metamorphism happens when a chunk of strata originally at the surface becomes deeply buried and subjected to squeezing horizontal stresses. When this happens, the sedimentary rock cracks, buckles, and is folded gently or severely depending on the amount of ongoing pressure. As the folds are shoved further down, heating increases and crystals begin to form as the sedimentary rock is changed into metamorphic rock. The speed and length of sedimentary burial affects the temperature and pressure it sees. For example, if the sediment is pushed down quickly in a subduction zone, it doesn't have time to heat up because of the high-pressure environment. However, if the downward movement is slow, the temperatures usually keep pace with the surrounding rock and mineral formation is slower, more complete, and gradual.

Regional metamorphism affects large structures across a broad stroke of the landscape. It involves the uplifting and down warping of stressed and deformed landmasses in the middle of mountain building. When both pressure and temperature increases are involved in regional metamorphism, it is called *dynamothermal metamorphism.*

Since regional metamorphism covers a large geographical area, the minerals and textures throughout the area are found in zones. Some areas may be near magma intrusion sources and contain zones of metamorphic and igneous rock. Some fairly undisturbed areas will look very different than those found nearer active tectonic areas. The main thing to remember is that in a broad region of metamorphism, the areas of changed rock can be found in horizontal and vertical positions.

Regional metamorphism produces rocks such as *gneiss* and *schist*. Regional metamorphism is caused by large geologic processes such as mountain building. These rocks, when exposed to the surface, show the unbelievable pressure that causes rocks to be bent and broken during the mountain uplifting process.

Schist rocks are metamorphic in origin. In other words, they started out as something else and were changed by external factors. Schists can be formed from basalt, an igneous rock; shale, a sedimentary rock; or slate, a metamorphic rock. Through tremendous heat and pressure, these rocks were transformed into this new kind of rock.

Schist is a medium-grade metamorphic rock. Medium-grade rock has been subjected to more heat and pressure than another rock such as slate. Slate, a low-grade metamorphic rock, needs lower temperatures for metamorphic changes to take place.

Schist is a coarse-grained rock with easily seen individual mineral grains. Since it has been compressed tighter than slate, schist is often found folded and crumpled. A lot of its original minerals have been transformed into larger flakes. Schists are usually named with reference to their original minerals. *Biotite mica schist, hornblende schist, garnet mica schist,* and *talc schist* are all different types of schist that come from different original minerals.

Gneiss rocks are also metamorphic in origin. Some gneiss rocks started out as granite, an igneous rock, but are changed by heat and pressure. Many gneiss rock samples have flattened mineral grains that have been smoothed flat by extreme heat and pressure and are aligned in alternating horizontal patterns.

Gneiss is a high-grade metamorphic rock. It has been the focus of much more heat and pressure than schist. Gneiss, a coarser rock form than schist, has distinct and easily seen banding. This banding is made up of alternating layers of different minerals. Gneiss can be formed from sedimentary rock such as sandstone or shale, or it can be created from the metamorphism of the igneous rock, granite. Since gneiss can come from granite, the same minerals found in gneiss are also found in granite. Along with mica and quartz, *feldspar* is the most important mineral found in gneiss. Gneiss is often used as a paving and building stone due to its attractive banding.

Dynamic Metamorphism

Dynamic metamorphism also results from mountain building. Huge extremes of heat and pressure cause rocks to be bent, crinkled, smashed, compacted, and sheared. Metamorphic rocks are generally harder than sedimentary rocks because of their tough formation environment and are hard or harder than igneous rocks. They form the bases of many mountain chains and are exposed as outcrops only after short-lived outer rock layers have been

worn away. Metamorphic rocks discovered in mountainous regions today provide geologists with clues as to the location of ancient mountains on modern-day plains.

Geologists use these clues to figure out the temperatures that change different rock types into metamorphic rock. The crystal arrangement of different rock samples gives them a good idea as to the temperatures that the specific sample has been exposed to during its lifetime.

Retrograde Metamorphic Rock

Sometimes a rock type is changed into a high-grade rock at one point, then later exposed to low temperatures and changed to another type of rock. When this happens, it is known as *retrograde metamorphism*. Retro means to go backwards in development.

An easy way to think of it is to picture butter. When butter is heated, it melts and turns into a liquid. When the temperature cools, the butter, which has separated into slightly different forms, goes back into a solid state. Later, if the butter is left out and melts at room temperature, it will eventually sour and return to its basic components.

Sometimes, geologists find rock that has been through more than one change. This is usually seen during microscopic crystal examination or through chemical analyses.

Pressure

There are three main factors that cause pressure increases and the formation of metamorphic rocks. These are:

- the huge weight of overlying sedimentary layers,
- stresses caused by plates clashing during mountain building, and
- stresses caused by plates sliding past each other, like the shearing forces along the San Andreas fault (western United States).

Pressure or *stress* from tectonic processes or the weight of overlying rock causes changes in mineral texture. The two types of pressure that are applied to existing rock are *confining pressure* and *directed pressure*.

Confining pressure is an all around pressure. Like atmospheric pressure at the surface of the Earth, confining pressure is present within the mantle's depths. Extreme confining pressure changes a mineral's structure by squeezing its atoms tighter and tighter until new minerals with denser crystalline structures are formed.

Directed pressure happens in a specific direction. When extreme squeezing pressure is applied in one direction, it's like toothpaste in a tube, it is forced in one direction. When clashing plates are compressed, the force is applied in one direction. Since heat decreases a rock's strength, when pressure is applied in one direction, a lot of folding and deforming goes on when temperatures are high.

Depending on the type of stress applied to a rock, the minerals in metamorphic rock are squeezed, stretched, and rotated to line up in a specific direction. This is how directed pressure affects the size and shape of metamorphic rock minerals undergoing change by heat and stress.

For example, during recrystallization of micas, crystals grow within the planes of their sheet-silicate structures and align perpendicular to the directed pressure. Geologists use this type of metamorphic mineral to figure out the pressures that specific samples have been exposed to during their history.

Chemical Changes

Characteristics that cause chemical changes in rocks also add to the formation of metamorphic rocks. Very hot liquids and gases can, under extreme pressures, fill the pores of existing rocks. These liquids and gases cause chemical reactions to occur, and over time, change the chemical composition of the existing rock. Metamorphism can take place instantly as in rock shearing at plate boundaries or can take millions of years as in the slow cooling of deeply buried magma.

It is important to remember that the changes that go on in metamorphism are mostly in rock texture. The chemical composition of metamorphic rock is altered very little. The basic changes that do occur include the addition or loss of water and carbon dioxide. The biggest changes of metamorphic transformation, then, have to do with the *way* minerals are rearranged.

A chemical shift in the composition of metamorphic rock can also be changed by the addition or removal of different elements. This can happen as a result of the intrusion of magma bringing new minerals into contact with existing rock. Sometimes this can be seen through color changes in minerals of the same basic chemical composition.

When hot, mineral-rich water rise through magma, they carry a variety of elements. Some of these elements include sulfur, copper, sodium, potassium, silica, and zinc ions to name a few. These minerals come from magma and intruded rock, during the time that water is filtering upward through the crust. On this journey, they interact with other minerals and chemicals replacing some of their own minerals with others. This type of chemical interaction and substitution is called *metasomatism*. Metal deposits like copper and lead are formed in this way.

Index Minerals

A Scottish geologist, George Barrow, noticed that rocks having the same overall mineral make up (like shale) could be seen to go through a series of transformations throughout specific zones in a metamorphic region. He found that minerals in individual zones had specific mineral configurations. As he studied minerals across a zone, he found that when new metamorphic mineral configurations were created, it was predictable.

> The first appearance of **index minerals** marks the boundary of low- to high-grade metamorphic rock changes in a specific regional zone.

Barrow found that mineral (shale) configuration changes happened with regard to *index minerals*. These index minerals acted like milestones in the low- to high-grade metamorphic rock transformation process. Barrow found that the domino effect of metamorphism happened in the following series:

> **chlorite** ⇨ **biotite** ⇨ **garnet** ⇨ **staruolite** ⇨ **kyanite** ⇨ **sillimanite**
> *Low grade* → *High grade*

When Barrow and his team studied the geological maps of the Scottish Highlands, they were able to plot where certain minerals started and stopped. They marked the locations of certain minerals and called these connected places *isograds*.

> An **isograd** is a marker line on a map connecting different areas of certain minerals found in metamorphic rock.

Metamorphic Rock Textures

Metamorphic rocks are divided into two categories, *foliated* and *nonfoliated.* Foliate comes from the Latin work *folium* (meaning leaf) and describes thin mineral sheets, like pages in a book. Metamorphic minerals that align and form bands, like granite gneiss and biotite schist, are strongly banded or *foliated.* Figure 8-4 gives you an idea of how mineral grains line up (foliate) after metamorphism has taken place.

> When metamorphic mineral grains align parallel in the same plane and give rock a striped appearance, it is called **foliation** or **foliated rock**.

Initially, the weight of sedimentary rock strata keeps the sheet-like formation of minerals parallel to the bedding planes. As the mineral layers are buried deeper or compressed by tectonic stresses, however, folding and deformation take place. The sedimentary strata are shoved sideways and are no longer parallel to the original bedding. In fact, metamorphism changes the texture enough that when broken, the metamorphic rock breaks in the direction of the foliation not the original mineral's composition. Table 8-1 lists some of the main foliated and nonfoliated metamorphic rocks and their mineral compositions.

Foliates are made up of large concentrations of mica and chlorite. These minerals have very clear-cut cleavage. Foliated metamorphic rocks split along cleavage lines that are parallel to the alignment of the rock's minerals. For example, mica can be separated into thin, flat nearly transparent sheets.

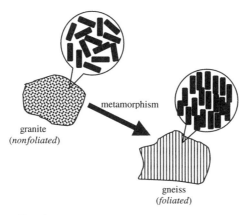

Fig. 8-4.   Foliated metamorphic rocks have parallel mineral grains.

Table 8-1 Metamorphic rocks are foliated and nonfoliated in composition.

Foliated or nonfoliated (f/n)	Metamorphic rock	Metamorphic grade (low, medium, high)	Mineral composition	Rock types
n	Marble	l → h	Calcite, dolomite	Limestone, dolostone
n	Quartzite	m → h	Quartz	Quartz sandstone
n	Amphibolite	m → h	Hornblende, plagioclase	Mafic igneous rock
n	Hornfels	l → m	Mica, garnet, andalusite, cordierite, quartz	Mudrock
f	Slate	l	Clay, mica, chlorite	Fine grain, splits easily
f	Schist	l → h	Mica, chlorite, quartz, talc, hornblende, garnet, staurolite, graphite	Mudrock, claystone, volcanic ash
f	Gneiss	h	Quartz, feldspars, hornblende, mica	Mudrock, sandstone, felsic, igneous rock

It is said to have good *schistosity*, from the Latin word *schistos* meaning easily cleaved.

> **Schistosity** is the parallel arrangement of coarse grains of sheet-structure minerals formed during the metamorphism and increasing pressure.

For fine-grained rocks with microscopic mineral grains, the breakage property is known as *rock cleavage* or *slaty cleavage*.

Slaty cleavage is found in an environment of low temperature and pressure. In these less-intense conditions, grain sizes increase and single grains are easily seen. Foliation is present with slaty cleavage, but not in a flat plane. Intermediate and high-grade metamorphic rock commonly breaks along rolling, or somewhat distorted surfaces along with the orientation of the grains of quartz, feldspar, and other minerals.

> **Rock cleavage** or **slaty cleavage** describes the way rock breaks into plate-like pieces along flat planes.

Large crystal textures can also be formed in a fine-grained, support rock during metamorphism. When this happens, crystals found in both contact and regional metamorphic rock are called *porphyroblasts*. They grow as the elements are rearranged by heat and temperature.

We learned that structural deformation goes on during metamorphism. When two rock surfaces deep in the Earth's crust grind against each other, crushing and stretching into bands, *myolites* are formed. These rocks are deformed under very high pressure. This deformation can take place before, during, or after metamorphic changes have happened and is part of the ongoing recycling of the rock.

For example, shale may be changed into schist during deep burial without any deformation. Then, much later, when tectonic action hauls the schist layer upward in mountain building, higher-grade metamorphism may cause foliation and deformation. Then, if the rock is living a really interesting life, it may be heated during contact metamorphism and change yet again.

Naming

When naming metamorphic rock, the rules are more flexible than that of igneous or sedimentary rock naming. Since metamorphic rock tends to change in composition and texture as temperatures and pressures change, the naming changes.

For example, shale is a fine-grained, clastic sedimentary rock containing quartz, clays, calcite, and some feldspar. With the start of low-grade metamorphism, *muscovite* and *chlorite* begins to form. Transformed shale is called *slate*. If the slate meets with further metamorphism, the mineral grains grow and intermediate-grade metamorphism happens with foliation and *mica* forms. Continued metamorphism causes the formation of even larger, coarse-grained rock with high schistosity and is known as *schist*. Then at high-grade metamorphism, the minerals group into separate bands with layers of mica-like minerals such as quartz and feldspar. This type of high-grade metamorphic rock is called *gneiss* from an old German word, *gneisto*, meaning to sparkle. Figure 8-5 shows the different types of metamorphic rock formed as temperature and pressure increase.

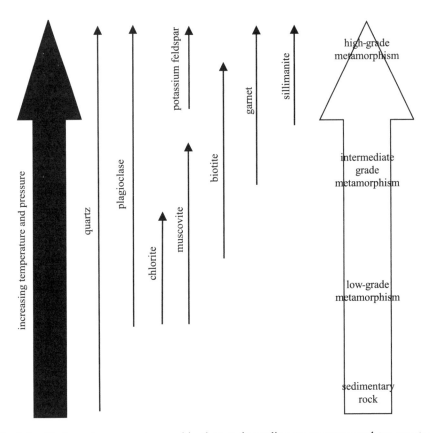

Fig. 8-5. Shale undergoes metamorphic change depending on pressure and temperature.

So naming then depends on what can be seen. Slate and phyllite describe textures, while gneiss is described by the large mineral grains (that are easily seen) being named first. So a specific gneiss might be named, quartz–plagioclase–biotite–garnet gneiss. In this way, another geologist would have a pretty good idea of all that the rock contained. Nongeologists would probably just call it garnet!

The Internet has several sites that provide photos of metamorphic rock types. There are even photos that illustrate the complete metamorphic rock series. I have listed several in the Reference section.

Quiz

1. Metamorphic rock is known as a
    (a) purple-colored rock
    (b) chameleon of rock types
    (c) surface only rock
    (d) ash rock type

2. Confining and directed pressure are
    (a) childbirth options
    (b) used to measure bicycle tire pressure
    (c) pressures measured in the core
    (d) two types of tectonic pressures

3. Transformed shale is called
    (a) slate
    (b) slant
    (c) sliver
    (d) magma

4. Metamorphic rocks are formed from
    (a) rocks that were originally something else
    (b) high-mountain rock
    (c) only igneous rock
    (d) only sedimentary rock

5. Slaty cleavage describes the way
    (a) some people dress in v-necked shirts
    (b) large sandwiches are cut at an angle
    (c) rock breaks into plate-like pieces along flat planes
    (d) rock forms in a volcanic vent

6. What marks the boundary of low- to high-grade metamorphic rock changes in a specific regional zone?
   (a) police crime scene tape
   (b) yellow curb paint
   (c) index minerals
   (d) transform faults

7. A marker line on a map connecting different areas of certain minerals is called a
   (a) thermocline
   (b) isograd
   (c) graded curve
   (d) isotherm

8. When mineral grains are parallel in the same plane, it is called
   (a) foliation
   (b) coloration
   (c) nonfoliation
   (d) deforestation

9. Metamorphic rock that contains both igneous and metamorphic rock is known as
   (a) chloride
   (b) shale
   (c) diatomite
   (d) migmatite

10. This metamorphic rock is often used as a paving and building stone
    (a) mica
    (b) gneiss
    (c) shale
    (d) obsidian

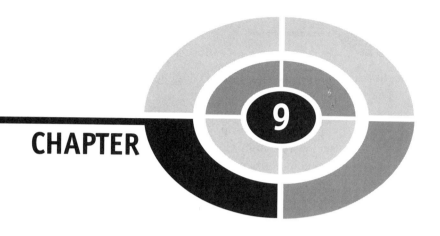

CHAPTER 9

Minerals and Gems

When you hear the word minerals, what comes to mind? Do you picture a cereal box advertising extra vitamins and minerals? Do you think of miners spending years searching for a glimpse of a shiny nugget or a brilliant stripe across a rock face? Or the many-faceted beauty of a friend's diamond ring?

In Chapters 6–8 we learned that rock found on the Earth's crust is a solid material created by three main geological processes: magma solidification, sedimentation of rock layers, and metamorphism. As a result, three basic rock types are formed.

- *Igneous rock* (volcanic or plutonic) is formed by the solidification of molten magma from the mantle.
- *Sedimentary rock* is formed from the burial, compaction, and lithification of deposited rock debris or surface sediments.
- *Metamorphic rock* is created when existing rock is chemically or physically modified by intense heat or pressure.

Geologists usually consider *rocks* to be a jumble of naturally occurring materials, mainly minerals. They can contain a mix of minerals and other organic substances ranging from microscopic mineral grains or organic matter to rough mineral agglomerates. Rocks can range in size from pebbles to mountains.

When people talk about their "rock collection," they usually mean their "mineral collection." Although some people collect rocks, mineral collectors are more common. They are the people looking for the "perfect" example of a specific mineral or the "rarest" specimen within a mineral group.

Amateur mineralogists and collectors are a lot like people who show dog breeds, like German Shepherds or English Pointers, to name a few. They get more points for having a specimen that meets the standard characteristics for the rock type and is of a high priority.

People value things that are rare and perfect. Flawless diamonds are much more valuable than those with flecks and flaws.

In fact, people have decorated themselves with shells, pieces of bone, teeth, and pebbles for the past 25,000 years (Paleolithic Period). But at that time most of the stones they chose were soft and brightly colored. Red carnelian and crystals were common choices.

From the time between 3000 and 2500 BC, *lapis lazuli* from Dadakshan reached Egypt and Sumer (Iraq). China, Greece, and Rome got their gemstones from many of the same regional mines.

Then, as people traveled and traded more, stones were made into family or governmental seals. They had different textures and some were carved. When they were rolled on damp clay, an imprint was made that identified a product. Seals were part of a leap in commercial trade. Some stone seals were worn around the neck and considered a status symbol. Kings and rulers had ring seals that were recognized as symbols of identity and power.

Ancient people thought gems and crystals had special powers. In an uncertain world, people wore them for protection. Color was important in the imagery. *Gold* was related to the Sun, blue to the sea, sky, or heavens, red to blood or the life force, and black for death. Wearing powerful gems was thought to protect the wearer's health, and bring wealth, luck, and love.

When the mummy of King Tutankhamun (1361–1352 BC) a Pharaoh of ancient Egypt, was discovered, it was decorated with gold, red carnelian, turquoise, crystals, jasper, obsidian, alabaster, amazonite, jade, and lapis lazuli. These were the amulets of wealth and strength at that time. These stones were worn during life and some placed on the mummy after death as a protection against harm in the afterlife.

Some minerals and gems were thought to be powerful by themselves, while others were thought to wield power through the figures and words written on them.

Minerals and gems were also thought to contain medicinal powers. The early Greeks recorded these claims in medical papers known as *lapidaries*.

The Greek philosopher, Theophrastus (372–287 BC) wrote the oldest surviving book on minerals and gems, called *On Stones*. He grouped 16 minerals into metals, earths, and stones (gemstones). A natural geologist, he accurately described physical characteristics of color, luster, transparency, hardness, fracture, weight, and medicinal benefits.

Pliny the Elder (AD 23–79) pulled together everything that earlier scholars had written into his 37-volume series, *Historia Naturalis*. Pliny's work provided a lot of useful information on sources, mining methods, uses, trade, and gem value.

Since the 1600s, scientists have become even more questioning. The study of minerals and gems has become a part of the study of chemistry, optics, and crystallography.

Minerals are often described by their chemical formulas in order to note the chemical substitutions of one or more atoms. For example, *topaz*, a prismatic crystal with the formula, $Al_2SiO_4(F_5OH)_2$, has been found to be as large as 100 kg. It can be colorless, white, gray, yellow, orange, brown, bluish, greenish, purple, or pink.

Gems and minerals are at the heart of the study of geology. Whether in the Earth or found on other planets, minerals tell the story of a planet's chemical and physical developments. They have specific characteristics with unique physical and chemical properties. This adds to their great variety and makes the study of minerals interesting.

The study of minerals, *minerology*, is usually focused on the external microscopic study of minerals in polished sections. People who hunt for and collect rough mineral specimens as a hobby are often called "rock hounds."

Mineral Groups and Properties

All minerals belong to a specific chemical group, which represents their affiliation with certain elements or compounds. The chemical structure of minerals is exact, or can vary slightly within limits. They have specific crystal-line structures and belong to different groups according to the way the mineral's atoms are arranged. Elements like *gold*, *silver*, and *copper* are found naturally and considered minerals.

> A **mineral** is a naturally found, inorganic substance with a specific crystalline structure.

Minerals are classified into the following chemical groups: elements, sulfides, oxides, halides, carbonates, nitrates, borates, sulfates, chromates, phosphates, arsenates, vanadates, tungstates, molybdates, and silicates. Some of these chemical groups have subcategories, which may be categorized in some mineral references as separate groups.

Nine Classes of Minerals

Geologists have identified over 3000 minerals. In order to study them more closely, geologists have divided minerals into nine different groups. Table 9-1 shows the broad groupings that minerals have been given.

Most rocks are composed of *minerals*. Minerals occur naturally as inorganic solids with a crystalline structure and distinct chemical make up. Table 9-2 gives you an idea of these different compounds. Minerals found in the rocks of the Earth are an assorted combination of different chemical elements. The most common elements that make up the minerals found in the Earth's rocks are illustrated in Fig. 9-1.

Table 9-1 Major mineral groups are determined by chemical composition.

Type	Chemical structure
1	Elements
2	Sulfides
3	Halides
4	Oxides and hydroxides
5	Nitrates, carbonates, borates
6	Sulfates
7	Chromates, molybdates, tungstates
8	Phosphates, arsenates, vanadates
9	Silicates

Table 9-2 Minerals are divided by different groupings.

Mineral	Group (element-e, halide-h, oxide-o, silicate-si, sulfide-su, phosphate-p, molybdate-m, borate-b, carbonate-c)	Hardness (*Mohs' scale*)	Chemical composition
Antimony	e	3–3.5	Sb
Arsenic	e	3.5	As
Bismuth	e	2–2.5	Bi
Carbon (diamond and graphite)	e	Graphite 1–2 Diamond 10	C
Copper	e	2.5–3	Cu
Gold	e	2.5–3	Au
Nickel–iron	e	4–5	Ni,Fe
Platinum	e	4–4.5	Pt
Silver	e	2.5–3	Ag
Sulfur	e	1.5–2.5	S
Fluorite	h	4	CaF_2
Halite	h	2.5	NaCl
Corundum	o	9	Al_2O_3 (ruby, sapphire)
Cuprite	o	3.5–4	Cu_2O
Hematite	o	5–6	Fe_2O_3
Albite	si	6–6.5	$NaAlSi_3O_8$

(*continued*)

Table 9-2   Continued.

Mineral	Group (*element-e, halide-h, oxide-o, silicate-si, sulfide-su, phosphate-p, molybdate-m, borate-b, carbonate-c*)	Hardness (*Mohs' scale*)	Chemical composition
Anorthite	si	6–6.5	$CaAl_2Si_2O_8$
Beryl	si	7–8	$Be_3Al_2(SiO_3)_6$
Dioptase	si	5	$CuSiO_2(OH)_2$
Jadeite	si	6–7	$Na(Al,Fe^{+3})Si_2O_6$
Labradorite	si	6–6.5	$(Na,Ca)Al_{1-2}Si_{3-2}O_8$
Microcline	si	6–6.5	$KAlSi_3O_8$
Olivine	si	6.5–7	$(Mg,Fe)_2SiO_4$
Orthoclase	si	6–6.5	$KAlSi_3O_8$
Quartz	si	2.65	SiO_2
Topaz	si	8	$Al_2SiO_4(F,OH)_2$
Zircon	si	7.5	$ZrSiO_4$
Cinnabar	su	2–2.5	HgS
Galena	su	2.5	PBS
Pyrite	su	6–6.5	FeS_2
Molybdenite	su	1–1.5	MoS_2
Gypsum	su	2	$CaSO_4\text{-}2(H_2O)$
Lazulite	p	5.5–6	$(Mg,Fe)Al_2(PO_4)_2(OH)_2$

(*continued*)

Table 9-2   Continued.

Mineral	Group (element-e, halide-h, oxide-o, silicate-si, sulfide-su, phosphate-p, molybdate-m, borate-b, carbonate-c)	Hardness (Mohs' scale)	Chemical composition
Turquoise	p	5–6	$CuAl_6(PO_4)_4(OH)_8 \cdot 4H_2O$
Wulfenite	m	2.5–3	$PbMoO_4$
Borax	b	2–2.5	$Na_2B_4O_5(OH)_4 \cdot 8H_2O$
Calcite	c	3	$CaCO_3$
Malachite	c	3.5–4	$Cu_2(CO_3)(OH)_2$
Rhodochrosite	c	3.5–4	$MnCO_3$

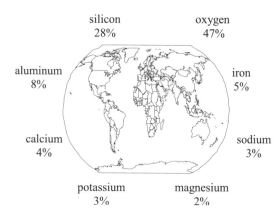

Fig. 9-1.   Earth minerals are composed of different elements.

ELEMENTS

The *elements* include more than 100 known minerals. Many of the minerals in this class are made up of only a single element. Geologists sometimes subdivide this group into metal and nonmetal categories. Of all of the elements, 80% are metals. Gold, silver, and copper are examples of metals.

Carbon produces the minerals *diamond* and *graphite*, which are nonmetals. Elements like phosphorus and selenium are also nonmetals.

For a complete listing of the known chemical elements, scientists use the *Periodic Table of Elements*. This is a chart that lists all the elements known today, along with a lot of other useful information. Besides the computer, the Periodic Table is probably the most important tool that scientists use.

Geologists use the Periodic Table to figure out the chemical composition of new minerals and to learn possible ways that different elements might bond.

The Periodic Table of Elements lists an element's symbol (shorthand name, like C for carbon, Al for aluminum), atomic number (equal to the number of protons), atomic weight, and sometimes the atomic energy levels of the element. When a certain element is described, it is written with the atomic number in superscript and the atomic weight in subscript. On a Periodic Table, *magnesium*, with atomic number 12 and an atomic weight 24.31, is written as:

While the simplest of Periodic Tables show just an element's atomic number and weight, complete charts give a broader amount of information. To give you an idea of the usefulness of the Periodic Table, the information listed for *titanium* in most Periodic Tables is shown below.

Ti
Titanium
Atomic Number – 22
Atomic Weight – 47.90
Group – 4
Period – 4
Transition Metal
Electrons per orbital layer – 2, 8, 10, 2
Valence electrons – 1s2 2s2p6 3s2p6d2 4s2

Knowing specifics about elements, like their electron arrangement, allows chemists and other scientists to figure out the bonding possibilities and types of compounds that can be formed with other elements. From this information, the mineral content of new and unknown samples is worked out. This information is also helpful when creating new compounds in the laboratory.

HALIDES

The *halides* are a group of nonmetals whose main chemical components include chlorine, fluorine, bromine, and iodine. Most halides are very soluble in water. They also form highly ordered molecular structures with a high degree of symmetry. *Halite* is the most common mineral of this group. It is known to most people as *rock salt*. Other halites include the minerals, *cryolite*, *atacamite*, *fluorite*, and *diabolite*.

OXIDES

A group of minerals, made up of one or more metals combined with oxygen, water, or hydroxyl (OH), is known as the *oxides* (and *hydroxides*) group. The minerals in this group show a great variety of physical characteristics compared to other more nonchanging groups. Some oxides are hard and others soft. Some have a metallic luster, while some are clear and transparent. Some of the oxide minerals include *anatase*, *corundum*, *chromite*, and *magnetite*, while hydroxides include *manganite*, *goethite*, *tungstite*, and *diaspore*.

SILICATES

The *silicates* encompass the largest minerals group. As the name implies, these minerals have a varying amounts of silicon and oxygen. Silicates are often opaque and light in weight. Silicate minerals are different from other groups in that they are all formed as *tetrahedrons*. However, it can be tough to identify individual minerals within the silicates group. A tetrahedron is a chemical structure where a silicon atom is bonded to four oxygen atoms (SiO_4). Some representative silicates include *albite*, *andesine*, *hornblende*, *microcline*, *labradorite*, *sodalite*, *leucite*, and *quartz*.

SULFIDES

The minerals of the *sulfide* group are often made up of a metal combined with sulfur and recognized by their metallic luster. The *sulfides* are an economically important group of minerals. The extraction of sulfide ores from composite metals is a standard process in industry. Specific ores are known for certain metal extractions, like *cinnabar* (a major source of mercury), *molybdenite* (molybdenum, an alloy in steel), *pyrite* (iron source), and *galena* (lead, used in piping and pewter).

SULFATES

The *sulfate* mineral group usually combines one or more metals with the sulfate compound, SO_4. Most sulfates are transparent to translucent, light in color, and soft. They usually have low densities. *Gypsum*, the most plentiful sulfate, is found in evaporite deposits. Common sulfates include *anhydrite* ($CaSO_4$) and *celestine* ($SrSO_4$).

Sometimes, sulfates contain substituted groups like *chromate*, *molybdate*, or *tungstate* in place of the sulfate group. *Chromates* are compounds in which metals combine with chromate (CrO_4). The minerals *crocoite* ($PbCrO_4$), *wulfenite* ($PbMoO_4$), and *scheelite* ($CaWO_4$) are all examples of different group replacements that form different minerals. These compounds are usually dense, brittle, and brightly colored.

PHOSPHATES

The mineral group, known as the *phosphates*, is made up of one or more metals chemically combined with the phosphate compound (PO_4). The phosphates are sometimes grouped together with the *arsenate*, *vanadate*, *tungstate*, and *molybdate* minerals. These minerals have substituted arsenic, tungsten, and molybdenum elements, respectively.

Although geologists list several hundred different types of these minerals, they are not common. *Apatite* is the most common phosphate mineral. Most minerals in these groups are soft, but their hardnesses can range from $1\frac{1}{2}$ to 5 or 6 (*turquoise*). Although brittle, they have well-formed crystals in beautiful colors like *lazulite* (blue) and *vanadinite* (red or orange).

CARBONATES

This is an easy one. *Carbonates* are minerals which contain one or more metals bonded with carbon in the compound (CO_4). Most pure carbonates are light colored and transparent. All carbonates are soft and brittle. They are usually found as well-formed rhombohedral crystals. Carbonates react with, bubble up, and dissolve easily in hydrochloric acid. Calcite is the most common carbonate. Other colorful carbonate minerals include *rhodochrosite* (pink to red), *smithsonite* (blue green), *azurite* (deep blue), and *malachite* (medium to dark green).

Nitrates and *borates* are often thought of as a subgroup of carbonates. They are formed when metal compounds combine with nitrogen and boron. When metals bond with nitrate, minerals like *nitratine*, a rare rhombohedral, transparent, often twinned mineral is formed.

When metals bond with borate, minerals like *borax*, *kernite*, and *ulexite* are formed. Most people have seen white borax, but it can also be colorless, gray, greenish, or bluish. Borax forms near hot springs, in ancient inland lakes, and places from which water has evaporated.

ORGANIC MINERALS

Minerals originally from organic sources (plants) are not usually classified as true (pure) minerals. However, some crystalline organic substances look and act like true minerals. These substances, formed primarily from carbon, are called *organic minerals*. *Amber* (petrified tree sap) is an example of an organic mineral.

Crystalline Structure

Most minerals can be found in crystalline form. Whether they are rounded, have shear faces, or have no set form, most minerals have specific internal geometric structures. Sometimes the structures are the same between different minerals, but their chemical compositions are different.

As individual as people, what goes on inside a mineral determines its physical and optical properties, shape, hardness, cleavage, fracture lines, specific gravity, refractive index, and optical axes. The regularly occurring arrangement of minerals, atoms, and molecules in space determines its form.

> The **lattice** structure of a mineral is based on its arrangement of atoms, ions, and molecules within an individual sample.

There are four different types of bonding that occur in crystalline solids. These determine what type of solid it is. The four types of crystalline solids are *molecular*, *metallic*, *ionic*, and *covalent*.

BONDING

These types of crystalline solids have molecules at the corners of the lattice instead of individual ions. They are softer, less reactive, have weaker non-polar ion attractions, and lower melting points.

A *molecular* solid is held together by *intermolecular* forces. The bonding of hydrogen and oxygen in frozen water shows how hydrogen forms bonds between different water molecules.

Another type of crystalline solid is made up of metals. All metals, except mercury, are solid at room temperature. The temperature needed to break the bonds between positive metal ions in specific lattice positions, like iron disulfide (FeS_2), and the electrons around them is fairly high. This strong bonding gives stable molecules flexibility. It allows metals to be formed into sheets (malleable) and pulled into strands (ductile) without breaking.

A metallic solid like silver is held together by a positively charged "central core" of atoms surrounded by a general pool of negatively charged electrons. This is known as *metallic bonding*. This arrangement of (+) ions and electrons (−) make metals good conductors of electricity.

Ionic solids form a lattice with the outside positions filled by ions instead of larger molecules. These are the "opposites attract" solids. The contrasting forces give these hard, ionic solids (like *magnetite* and *malachite*) high-melting points and cause them to be brittle. Hardness is not the same as *brittleness*. Brittleness, a measure of mineral strength, is dependent on a mineral's overall structure. Think of it like building a house without the proper internal supports. Brittle minerals fracture easily. Figure 9-2 shows the way crystalline solids can be arranged.

Ionic bonding in a solid occurs when anions (−) and cations (+) are held together by the electrical pull of opposite charges. This electrical magnetism is found in a lot of salts like potassium chloride (KCl), calcium chloride (CaCl), and zinc sulfide (ZnS). Ionic crystals, which contain ions of two or more elements, form three-dimensional crystal structures held together by the

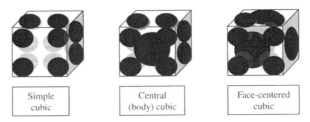

Fig. 9-2. Crystalline solids have different configurations.

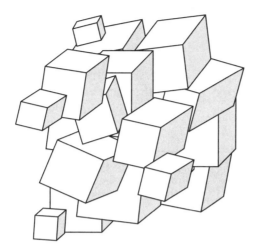

Fig. 9-3.   Sodium chloride has a cubic crystalline structure.

strong ionic bonds. Figure 9-3 illustrates the cubic arrangement of table salt (halite, NaCl).

Covalent bonding holds hard solids together. Assembled together in large nets or chains, covalent multilayered solids are extremely hard and stable in this type of configuration. Diamond atoms use this type of structure when arranged into three-dimensional solids. One carbon atom is covalently bonded to four other carbons. This strong crystalline structure makes diamond the hardest known organic solid.

Covalent crystals are all held together by single covalent bonds. This type of stable bonding produces high melting and boiling points.

> *Allotropes* are different structural forms of the same element. Graphite, diamond, and buckminsterfullerene are all allotropes of carbon.

The different bonding and forms of carbon in a diamond (pyramid shaped), graphite (flat-layered sheets), or buckminsterfullerene (C_{60} and C_{70}, shaped like a soccer ball) illustrate the variety and stability of covalent molecules. Nets, chains, and balls of carbon bonded into stable molecules make these solids hard and stable.

Minerals also have well-studied properties, such as color, hardness, crystalline structure, specific gravity, luster (shine or luminescence), cleavage, and tensile strength (resistance to being pulled apart). Many of these properties can vary slightly within a single mineral. Some minerals have very specialized properties like fluorescence and radioactivity.

Habit

Minerals come in many different sizes, shapes, and colors. The diversity and combination of colors within the same chemical formula keeps mineralogists guessing when they collect a new sample that doesn't seem to fit the system.

> A mineral or aggregate's physical size and shape are called its **habit**.

There are several basic mineral habits mostly used to identify mineral specimens. They include the following:

- *Acicular* (thin, needle-like masses),
- *Bladed* (sharp-edged, like a knife),
- *Dendritic* (plant-like shape),
- *Fibrous* (furry),
- *Granular* (grainy),
- *Lamellar* (thin layers, plates, or scales),
- *Massive* (no specific shape),
- *Reniform* (rounded, globular masses),
- *Rosette* or *radiating*,
- *Prismatic* (flat or pointed ends with long, parallel sides), and
- *Tabular* (overlying flat squares).

A sampling of minerals with different habit types are listed in Table 9-3. Depending on the conditions present at the time crystals are formed, broad differences in a mineral or aggregate's habit are possible.

TWINNING

When a mineral sample has two or more nonparallel crystals that intersect and grow together, it is known as *twinning*. Twinning is often found in twin sets. A rare *chrysoberyl* specimen, measuring 8 cm across and containing three twinned crystal sets, was found in Espirito Santo, Brazil. This is an example of a chrysoberyl *trilling*.

When the crystals push against each other and form a mass, it is called *contact twinning*. However, if one penetrates and cuts through the structure of another at an angle, it is known as *penetration twinning*.

Table 9-3 A mineral's habit is a good visual way to identify it.

Mineral	Habit
Actinolite	Bladed
Alabaster	Granular, massive
Anhydrite	Fibrous, granular, massive
Barite	Prismatic
Beryl	Prismatic
Chalcanthite	Tabular, fibrous
Copper	Dendritic
Gypsum	Rosette, radiating
Hematite	Reniform
Limonite	Massive
Mica	Lamellar
Scolecite	Acicular
Titanite	Massive, lamellar
Silver	Dendritic
Zircon	Prismatic

CLEAVAGE

In geology, *cleavage* is determined by the way a mineral breaks when struck with a rock hammer. Depending on the crystalline structure, it cleaves between flat, well-defined planes. These planes are separated between layers of atoms or other places, where bonding between atoms is weakest. Cleavage faces are not as smooth as crystalline faces, but tend to cleave the same way each time the sample is broken. Depending on the structure of the mineral, cleavage breaks are described as *perfect* (breaks along the base or between crystals in the sample), *distinct*, *indistinct*, or none. Most minerals with *basal*, *rhombic*, *prismatic*, or *cubic* cleavage break along or between parallel planes. Those mineral types are commonly large and easy to spot. Galena, dioptase, and hematite are all examples of minerals with crystalline structures that break along cleavage planes.

FRACTURE

When you hit a sample with a rock hammer and it breaks without any real rhyme or reason, this is called a *fracture*. The sample has surfaces that are rough and uneven (compared to the easily seen shapes of cleaved samples). Most minerals fracture and cleave depending on their habit, but some only fracture. Fractures are described as *uneven*, *conchoidal* (shell-like), *jagged*, and *splintery*. A rough opal, for example, splits into a curved, shell-like fracture. The different parts of the split can have a wide spectrum of colors, from light blue to the rainbow of color found in "fire opal."

HARDNESS

A physical characteristic of mineral identification that doesn't change from one sample to another is *hardness*. Hardness is constant because a mineral's chemistry is usually constant. Samples of the same mineral content can change a bit from one to the next, but in general they are about the same. Variations are only found when a mineral is poorly crystallized or is really an aggregate of different minerals.

Minerals with tightly packed atoms and strong covalent bonds are the hardest minerals. Minerals with metallic bonds or weak interconnected forces are the softest minerals. *Talc*, rated at the bottom of the hardness scale, is an example of an extremely soft mineral.

> A mineral's **hardness**, established by its physical structure and chemical bonding, is its resistance to being scratched.

Hardness is tested through scratching. A scratch on a mineral is actually a mark produced by surface microfractures of the mineral. Fractures take place when bonds are broken or atoms are pushed aside (metals). A mineral can only be scratched by a harder mineral.

In 1812, French mineralogist, Friedrich Mohs, proposed a scale using set values as standards to test an unknown sample's hardness against. Before Mohs set the standard, hardness was mostly done through guesswork. It was tough to describe hardness to other geologists unless they were right there in the field or lab holding the sample themselves.

> The **Mohs' Scale of Hardness** starts with talc at 1 and ends with diamond at 10, the higher the number, the harder the mineral.

This scale is not precise, but it gives geologists a common frame of reference to use when testing a sample's hardness. Table 9-4 shows the relative hardness of minerals on the Mohs' Scale of Hardness. To give you an idea of how common items compare in hardness to the Mohs' scale, several common things are listed with their Mohs values.

The Mohs' Hardness scale is one tool used by geologists and mineralogists around the world to tell different minerals apart. To use this scale, it is necessary to have some of the minerals found in the scale on hand.

Table 9-4   Mohs' Scale of Hardness is used to test the comparative hardness of samples.

Mineral hardness (Mohs' scale)	Mineral	Common hard stuff
1	Talc	Pencil lead (1–2)
2	Gypsum	Fingernail ($2\frac{1}{2}$)
3	Calcite	Penny
4	Fluorite	
5	Apatite	Knife blade ($5\frac{1}{2}$)
6	Orthoclase	Glass (6)
7	Quartz	Garnet
8	Topaz	
9	Corundum	Ruby, sapphire
10	Diamond	

Some geologists begin hardness testing of an unknown mineral against *orthoclase* to see if the unknown mineral can scratch it. If the unknown mineral scratches the orthoclase, then it must be of hardness greater than 6. If the *apatite* scratches the unknown, then the unknown mineral must be of a hardness less than 6. If they scratch each other, then the unknown sample has a hardness of 6.

To get closer to an unknown mineral's hardness, it can be tested against other less hard standards like apatite or fluorite. If it is softer than apatite and fluorite, try gypsum until you find the approximate hardness. Since the Mohs' scale is a relative scale, one mineral sample may be scratched by another and be given a certain hardness, but it might be slightly more or less depending on other factors like shape or size.

It is important to remember to perform a hardness test on the backside or not easily seen part of a mineral. Some inexperienced collectors and students, in their excitement to discover more about a mineral, scratch right across a perfect crystal face. This ruins the specimen for display or jewelry! A fractured, cleaved, or unnoticeable part of the mineral still gives an accurate hardness test and doesn't damage a beautiful specimen's best face.

If they don't have a Mohs' Hardness Scale, some amateur geologists and students add a "hardness kit" to their rock hunting gear. The Mohs' scale is used for wide comparisons between minerals, so testing a sample with a fingernail, copper penny, or knife blade often gives a rough idea as to its hardness.

Table 9-5 gives you a few hardness hints to look for when testing different mineral samples for hardness. One way to remember the minerals on the Mohs' scale is to make up a memory aid using the first letter of each of the Mohs' minerals (*t*alc, *g*ypsum, *c*alcite, *f*luorite, *a*patite, *o*rthoclase, *q*uartz, *t*opaz, *c*orundum, and *d*iamond). It can be anything. Mine is, "*T*he *G*eologist's *C*at *F*ound *A*n *O*ld *Q*ueen's *T*offee *C*olored *D*iamond."

Remember that the Mohs' Scale of Hardness is comparative and not absolute. Fluorite, with a hardness of 4, is not twice as hard as gypsum with a hardness of 2. Although talc is a 1 and diamond a 10 on the Mohs' scale, the hardness difference between them is really about one hundred fold. The hardness differences between calcite and fluorite (3 and 4) are not the same as the differences between corundum (9, like ruby and sapphire) and diamond (10).

Hardness is especially important when choosing gemstones. Except for apatite (5), turquoise (5–6), and opal ($5\frac{1}{2}$–$6\frac{1}{2}$), very few soft minerals can be cut as gems. People with jewelry made from these minerals are usually warned against cleaning them in vibrating cleaning machines since they can easily break.

Table 9-5   When testing for hardness, check for these common characteristics.

Characteristic	Hardness hints
Orientation	Most minerals have small differences in hardness depending on the direction and orientation of the scratch. Kyanite samples have a hardness range of $(5\frac{1}{2} - 7)$
Size	A 1500 kg specimen is often softer than a single crystal because of the crystal structure. Hardness is truer when tested on individual crystals
Purity	Some minerals have a range of hardness values because of impurities or ion substitution
Dust	Sometimes there is a dust trail on a harder, unaffected mineral after being *"scratched"* by a softer mineral. Always blow or rub across a scratch to be sure there is a real scratch
Scratching ease	Relative hardness is affected by ease of scratching. Both diamond and quartz scratch glass, but diamond scratches glass with extreme ease, like a knife through butter

Soft minerals are usually best for viewing and not for wearable jewelry. People who buy malachite $(3\frac{1}{2} - 4)$ earrings and drop one on a hard surface are surprised when it shatters. After all, their amethyst (7) earring hadn't broken when it was dropped. Common gemstones like topaz (8), jasper (7), and aquamarine (7–8) have a hardness of 7 or more. Hardness also plays a big part in the selection of industrial minerals used for grinding, polishing, and other abrasive tasks. Soft minerals like talc and graphite are used as high-temperature lubricants, pencil lead, talcum powder, and to give shine to paper.

An *absolute hardness* scale has different values than the relative Mohs' scale. Using precise instrumentation, mineralogists are able to measure the absolute hardness of minerals with much more precision. Most minerals are fairly close in hardness, but as hardness increases, the hardness differences increase by greater and greater amounts. Table 9-6 shows a comparison between the absolute hardness values and the relative Mohs' hardness values for the same minerals.

> **Absolute hardness** is a precise measurement of a mineral's hardness and not dependent on a comparison with other samples.

Table 9-6   There is a big difference between the Mohs' relative hardness scale and absolute hardness.

Mineral	Mineral hardness (*Mohs' scale*)	Mineral hardness (*Absolute scale*)
Talc	1	1
Gypsum	2	3
Calcite	3	9
Fluorite	4	21
Apatite	5	48
Orthoclase	6	72
Quartz	7	100
Topaz	8	200
Corundum	9	400
Diamond	10	1600

For example, the absolute hardness of talc is 1. Diamond is 1600 times harder! When most people talk about diamonds, rubies, and sapphires, they consider them to be the same hardness and lump them together. However, geologists know better. Rubies and sapphires are different varieties of corundum which has an absolute value of 400. Diamonds are four times harder with an absolute value of 1600.

It's easy to see why diamond gets a lot of respect as the Earth's hardest natural mineral. Although there are a lot of compounds being formed and studied with the idea of creating something harder than diamond, the super-compressed, tightly bonded structure of carbon (diamond) is pretty amazing.

Most minerals have small differences in hardness according to the direction of the scratch and the orientation of the scratch. The environment in which a mineral formed within a rock can affect its hardness. For example, *cyanide* has a range $(5\frac{1}{2} - 7)$ of hardness levels depending on these factors. Impurities and ion substitutions can also affect the hardness of a sample.

A huge specimen (several hundred pounds) is often softer than a single crystal because of its crystal structure, so hardness is most accurate when tested on individual crystals.

Sometimes a *dust trail* appears on a mineral after it has been *"scratched"* by a softer mineral. It looks as if the softer mineral has scratched the harder mineral, but the "scratch" is really just a dust trail across the unyielding surface of the harder mineral.

COLOR

Most people are most interested in the *color* of a mineral. This is especially important for choosing minerals as gemstones for jewelry. After all, jewelry has to match the outfit (maybe that's why "diamonds are a girl's best friend," they go with everything)!

One of the problems geologists find in using color to identify minerals of a certain group is that some minerals can be very different.

Some colors are called *idiochromatic*. Their chemistry gives them their color. Malachite which has a lot of copper is always green because copper gives it that color.

Minerals that are usually colorless and take on the color of small impurities are called *pseudochromatic*. Depending on the impurity, they can have a variety of colors. If a mineral contains bits of iron, it will take on a reddish color.

Allochromatic minerals are generally colorless and transparent. They get their color from the small changes in their crystalline make up or from structural flaws. In corundum, for example, the substitution of iron and titanium for aluminum gives blue sapphire, while iron by itself produces a yellow sapphire.

Minerals like *quartz* come in lots of colors. Some of the colors that quartz can take are listed below:

- Colorless quartz,
- Rose quartz (all shades of pink),
- Milky quartz (white and whitish gray),
- Citrine quartz (yellow, yellowish brown, and orange),
- Smoky quartz (brown, brownish black, and black), and
- Amethyst (light to deep purple).

Many minerals, depending on their chemical content and formation, are found in different colors. Ruby and sapphire are both varieties of corundum with the same chemical composition (AlO_3) and hardness (9), but two very different colors. Most people know that rubies are red and sapphires are blue. However, just to keep you guessing, sapphires can also be colorless, green, yellow, or purple!

Tourmaline is thought to have the greatest number of color variations. A "chameleon of color," tourmaline is a prismatic crystal and is found to occur in seven different forms. These include, *elbaite* (multicolored), *schorl* (black), *buergerite* and *dravite* (brown), *rubellite* (pink), *chromdravite* (green), and *uvite* (black, brown, and yellowish green). Long, tourmaline crystals can be pink on one end and green on the other! They look like some rare cosmic gem from a science fiction movie.

A few minerals, like ruby, are fluorescent. They absorb blue and ultraviolet light and then release some of the energy back in the red part of the light spectrum.

STREAK

The *streak* of a mineral is simple to remember. It is just what it says, a powdery streak made when rubbing a sample across an unglazed surface. Streak is a more dependable way to test a mineral than color since it is nearly always the same for different minerals. Sometimes a hard sample must have a small bit crushed with a geological hammer to get a sample to test.

A streak may be colorless, white, golden yellow, yellow, reddish brown, red, gray, brown, or black. The streak of a mineral is often not the same color as the mineral appears to the eye. For example, the mineral *crocoite* is orange–red and its streak is yellow. Wulfenite can be orange, yellow, brown, gray, or greenish brown, but its streak is white. Table 9-7 shows a variety of minerals with their visible color, streak, and luster.

LUSTER

Luster is the word geologists use to describe the way light reflects off the surface of a mineral or crystal. The amount of light absorbed and a mineral's texture affect luster. The different types of luster consist of dull, metallic, *vitreous* (glassy), *adamantine*, pearly, greasy, silky, and waxy. These are pretty straightforward and were used by some of the earliest people in describing different minerals. For example, *gold* and *platinum* have metallic lusters, but not *microcline*, which has a vitreous or pearly luster. Most silicates, sulfates, halides, oxides, hydroxides, carbonates, and phosphates have a vitreous luster.

Table 9-7   A mineral's streak can be a different color from the mineral itself.

Mineral	Visible color	Streak	Luster
Albite	Colorless, white, bluish, gray, greenish, or reddish	White	Pearly to vitreous
Antimony	Silvery-gray	Gray	Brilliant metallic
Arsenic	Pale gray, dark gray	Pale gray	Metallic
Bismuth	Silvery-white (reddish tarnish)	Silvery-white	Metallic
Cinnabar	Brownish-red or scarlet	Scarlet	Adamantine, submetallic, or dull
Copper	Rose-red, copper-red	Copper-red	Metallic
Corundum	Many colors, red (ruby), blue (sapphire)	White	Vitreous to adamantine
Cuprite	Red	Brownish-red	Adamantine, submetallic, or earthy
Diamond	Colorless, white, gray, orange, yellow, brown, pink, red, blue, green, or black	White	Adamantine or greasy
Dioptase	Deep bluish-green	Pale bluish-green	Vitreous
Erythrite	Deep purple to pale pink	Purple to pale pink	Adamantine, vitreous, or silky

(continued)

Table 9-7 Continued.

Mineral	Visible color	Streak	Luster
Fluorite	Colorless, purple, green, white, yellow, pink, red, blue, and black	White	Vitreous
Galena	Lead-gray	Lead-gray	Metallic
Gold	Yellow	Golden yellow	Metallic
Graphite	Dark gray to black	Dark gray to black	Dull metallic
Hematite	Bright red, brownish-red, steel-gray, iron-black	Brownish-red	Metallic to dull
Jadeite	Green, white, gray, or mauve	Colorless	Vitreous to greasy
Labradorite	Colorless, blue, gray, or white	White	Vitreous
Lazulite	Blue, bluish-green	White	Vitreous to dull
Malachite	Deep green	Pale green	Vitreous to silky
Microcline	Green, white, gray, yellowish, reddish, or pink	White	Pearly or vitreous
Molybdenite	Gray	Gray	Metallic

Mineral	Color	Streak	Luster
Nickel–iron	Steel-gray, dark gray, or blackish	Steel-gray	
Olivenite	Olive-green, brown, yellowish, gray, or white	Olive-green	Vitreous to silky
Olivine	Green, greenish-yellow, white, or brown	Colorless	Vitreous
Platinum	Silvery-gray to white	White, silvery-gray	
Pyrite	Pale yellow	Greenish-black	Metallic
Rhodochrosite	Pink to red	White	Vitreous to pearly
Silver	Silvery-white	Silvery-white	Metallic
Sulfur	Lemon-yellow to yellowish-brown	White	Resinous to greasy, adamantine
Wulfenite	Brownish-black	Reddish-brown to black	Submetallic

A diamond's high luster or that of highly reflective, transparent, or translucent minerals is known as an *adamantine* luster. *Zircon, cuprite,* and some forms of *sulfur* and *cinnabar* have this type of luster.

One thing to remember is that depending on the mineral and the environment in which it was formed, luster can be different in different parts of the same sample, as well as in different samples (from different places) of the same mineral.

TRANSPARENCY

Depending on the way minerals are bonded, light will pass through a mineral in different amounts. When you can see right through a mineral like glass, it is said to be *transparent*. If light is slightly blocked, making the mineral look foggy and unclear, it is said to be *translucent*. If a mineral sample is solid and lets no light pass through at all, it is called *opaque*.

> A **transparent** mineral can be seen through, while a **translucent** mineral is hazy, and an **opaque** mineral lets no light pass through at all.

A sample's transparency isn't always the same all the way through the sample. Crystals are often transparent to translucent across a sample. For example, *amethyst* and *olivine* crystals are usually transparent to translucent in a sample. *Opal* is transparent to opaque across a sample, while *copper* and *jamesonite* are opaque.

SPECIFIC GRAVITY

When geologists are trying to figure out the identity of an unknown sample, they use the above-mentioned characteristics as well as *specific gravity* (SG). The density of a sample is measured in terms of its specific gravity.

> **Specific gravity** is the ratio of the mass of a substance compared to the mass of an equal volume of water at a specific temperature.

To find the specific gravity of a mineral, compare its weight to the weight of an equal volume of water. For example, a specific gravity of 4 tells geologists that an unknown sample is four times heavier than water. Size doesn't matter. A larger sample can have a lower specific gravity. This is the case with *talc* and *mercury*. A large amount of talc would have a lower specific gravity than a small amount of mercury. The specific gravity of talc is 2.8, while *mercury*'s specific gravity is 13.6. Table 9-8 gives some common minerals and their specific gravities.

Gold, Silver, and Copper

Since the first shiny speck caught the eye of early humans, gold, silver, and copper have been used for coins, jewelry, and household serving ware. Resistant to rust and corrosion, they were an excellent choice for coins, while being doubly useful in showing the local king's face to strangers passing through the country.

Gold, a shiny yellow metal, is a good conductor of heat and electricity. It is the most malleable and ductile metal. The early alchemists based their reputations and lives on providing more of this metal to their patrons.

In the western United States during the Gold Rush days of the 1800s, *gold fever* affected thousands of people seeking their fortunes and a better life. Miners spent from morning until night hunched over icy mountain streams panning stream gravel and watching for the bright glint of a single gold nugget. Even today, people get excited over flecks of gold (iron sulfide, FeS_2) found in rock.

Silver, a brilliant white, lustrous metal, is the best conductor of heat and electricity of all the metals. It was also prized by early peoples for its beauty and uses. Silver, though, is less resistant to corrosion and will tarnish, turning black when it oxidizes in the air. It was thought that the state of Nevada was admitted to the Union in 1864 during the Civil War to provide funds to the Union and easier access to its resources of silver. Silver is used in coins, jewelry, electrical contacts, mirrors, circuitry, photography, and batteries.

Copper has an orange-brown color that is used in pipes, electrical wires, coins, paints, fungicides, and in alloys combined with other metals. In some countries, local artisans use copper for platters, bowls, tools, and jewelry. Pennies though once 100% copper are now (since 1981) only treated on the outside with copper plating to give the United States one cent coin its reddish-brown (copper) color. Many years ago, the badges of policemen were made from copper and so the slang expression "copper" or "cop" was used.

Table 9-8   Minerals of different densities have different specific gravity values.

Mineral	Specific gravity
Albite	2.61
Antimony	6.7
Arsenic	5.7
Cinnabar	8.09
Copper	8.95
Corundum	4.0–4.1
Diamond	3.5–3.53
Fluorite	3.18
Galena	7.58
Gold	15–19
Graphite	2.09–2.23
Hematite	5.26
Jadeite	3.33
Labradorite	2.7
Lazulite	2.4
Magnetite	5.18
Malachite	4.05
Manganite	4.33
Marcasite	4.89
Mercury	13.6

(continued)

Table 9-8   Continued.

Mineral	Specific gravity
Molybdenite	4.6–4.7
Olivine	3.2–3.4
Opal	2.1
Platinum	14–19
Pyrite	5.02
Scheelite	6.10
Silver	10–11
Spinel	3.55
Sulfur	2.07
Talc	2.8
Topaz	3.5–3.6
Zircon	4.7

Ores

Unlike gold and silver, which are pure elements, many metals are not found in nature as a single element. Most metals are combined with other elements within ores that must be processed to extract their different parts. Table 9-9 gives examples of different ores and the metals they contain.

Pure metals are separated from ores primarily by heat. This is done in a high-temperature blast furnace. By adding reactants, like limestone and coke (a carbon residue) to break hydrogen bonding and release the bonded metals, individual metals can be collected.

Lead, though sometimes found as a pure metal in nature, is usually found as the ore galena or lead sulfide. Lead ore is crushed, heated in a blast furnace, and then extracted. Most lead produced in the United States is used

Table 9-9   Mineral-rich ores contain a combination of two or more elements.

Element	Ore	Found in
Aluminum	Bauxite	France, Jamaica
Bismuth	Bismite	USA
Chromium	Chromite	South Africa, Russia
Cobalt	Cobaltite	Germany, Egypt
Copper	Chalcopyrite	Cyprus, USA, Canada
Iron	Hematite	USA, Australia
Lead	Galena	USA, Brazil, Canada
Mercury	Cinnabar	Algeria, Spain
Nickel	Pentlandite	Canada
Tin	Cassiterite	Bolivia
Tungsten	Wolframite, Scheelite	Spain, China
Zinc	Sphalerite	Australia, Canada, USA

for batteries, and batteries' electrodes as well as lead solder are used in making connections on computer circuit boards.

Mercury is most often found in nature as the ore, *cinnabar*. Cinnabar, also called *vermillion*, is a bright red mineral that was crushed to a powder and used by Renaissance painters to make a deep red paint pigment.

GEODES

Have you ever found a round, average looking rock and cracked it open to find beautiful crystals inside? It's like finding hidden treasure. Round or oval rocks that have an open center are called *geodes*. The word *geode* comes from the Greek, *geoides*, which means "earthlike."

> A **geode** is a sphere or fairly round rock that contains a hollow open space lined with crystals.

Some geodes were originally: (a) air trapped in volcanic rock or openings, (b) balls of mud, or (c) plant material trapped in sedimentary rock. Eventually, the outer geode skin hardened and water-containing silica seeped into the inside walls of the geode's hollow cavity. Over geological time, crystals of *quartz*, *calcite*, *amethyst*, and others were formed from the evaporation and reaction of minerals that were in the water. Depending on the geode's silica content, different crystal layers are created.

Geodes that are completely filled with small crystal formations such as *quartz*, *agate*, *jasper*, or *chalcedony* are called *nodules*. The only difference between a geode and a nodule is that a geode has a hollow cavity and a nodule is solid.

Quartz is the most common mineral found in many geodes, but *kaolinite*, *calcite*, *pyrite*, and *sphalerite* are also found. Depending on prehistoric environmental conditions, like fluctuations in sea level and sedimentation pressure, crystals of *aragonite*, *barite*, *chalcopyrite*, *dolomite*, *goethite*, and *marcasite* were formed in geodes.

METEORITES

Geologists sometimes find rocks that have come to the Earth from other parts of the solar system. A few are boulder size, but most don't survive their fiery entrance through the atmosphere. Those that do are smaller, more like the size of a baseball.

Meteorites are classified in the following groups:

1. *Iron* – mostly nickel iron alloys (4–20% nickel),
2. *Stony-iron* – metals and silicates, and
3. *Stony* – mostly magnesium and iron silicates.

Some meteorites contain small aggregates of other matter. Mineral aggregates are called *chondrites*. Chondrites contain water-bearing minerals and carbon compounds are known as *carbonaceous chondrites*. This type of chondrite is thought to be of extremely ancient origin, perhaps even from the original formation of the solar system.

Many people are interested in minerals and gems besides geologists. Scientists, who study the development and dating of rock strata, spend long hours working in freezing alpine mountains or hot desert outcrops to collect

specimens for later analysis. Different mineral deposits can be found in other nearby outcrops depending on the mineral type, texture, and environment to which it has been exposed.

Minerals are important national assets. They are mined for valuable elements or an essential property they possess. Minerals, like platinum, are mined for their magnificence and rareness. There are roughly 3000 different types of minerals and new ones are found regularly. A lot of mineral types are unknown to mineral collectors, because they are rare, have no economic value, and are not particularly attractive.

However, as technology advances, new uses for previously unimportant minerals may be found.

Quiz

1. Geodes that are completely filled with small crystal formations are called
    (a) nodules
    (b) spheroids
    (c) noodles
    (d) phenocrysts

2. The ratio of the mass of a mineral compared to the mass of an equal volume of water at a specific temperature is called the
    (a) atmospheric pressure
    (b) specific gravity
    (c) lean mass
    (d) heavy water

3. A mineral that light can shine all the way through is said to be
    (a) transparent
    (b) translucent
    (c) opaque
    (d) priceless

4. Adamantine and metallic are types of
    (a) specific gravity
    (b) mantle
    (c) luster
    (d) habit

5. What mineral is most commonly found in geodes?
   (a) silver
   (b) quartz
   (c) hematite
   (d) tourmaline

6. On the absolute hardness scale, diamond is how many times harder than talc?
   (a) 1
   (b) 400
   (c) 800
   (d) 1600

7. The Periodic Table gives information about a mineral's
   (a) atomic number and mass
   (b) period
   (c) group
   (d) all of the above

8. A mineral is
   (a) a person who mines for a living
   (b) a liquid nitrogen sample
   (c) a homogeneous sample with a specific crystalline structure
   (d) always made up of two or more oxides

9. A powdery line made by rubbing a sample across an unglazed surface is called a
   (a) scratch
   (b) mess
   (c) streak
   (d) flaw

10. The precise measurement of the hardness of a mineral, not dependent on comparison with other samples, is called
    (a) complete stiffness
    (b) absolute hardness
    (c) total rock
    (d) Mohs' scale

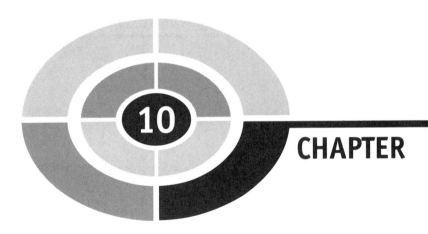

Fossils

CHAPTER 10

There are few places more interesting or awesome to visit than a museum of natural history. These institutions that preserve the past can open up a window on the world of the prehistoric Earth and its inhabitants. Dinosaurs, mammoths, ancient whales, and other extinct creatures are displayed for all to see. Fossil bones and replicas of immense, extinct creatures are amazing and scary at the same time. These creatures are the subject of wild imaginings as well as best-selling books and movies.

What is a fossil anyway? Are they really rocks? Where do you find fossils? How do fossils form?

In this chapter, we'll look at fossils with the view of how they fit into the Earth's geology and history. The books listed in the references provide a good start to the study of dinosaur types and lifetimes. For a complete study of dinosaurs, marine life, and other prehistoric creatures and fossils, the following Internet websites provide more information:

www.mnh.si.edu
www.nhm.ac.uk
www.nationalgeographic.com/features/96/dinoeggs

It has only been within the past 200 years that fossils were recognized for what they are. Before that, people thought strange-shaped rocks that looked like shells, wood, lizards, and other creatures were just oddities of nature, like a rock formation that looks like a bear or the face of a person.

Fossils

Since most people today read about and play with toy dinosaurs from childhood, it's hard to imagine a time when no one really knew what fossils were or how they were formed. Usually, depending on where they were discovered, creative explanations thrived. It was not until Darwin published, *On the Origin of Species by means of Natural Selection* in 1859 (a best seller at the time), that the idea of extinct ancestors of modern plants and animals seemed possible.

The field of *paleontology* began about this time. A *paleontologist* is a scientist who studies fossils. Paleontology includes all fossils, not just dinosaur bones.

A fossil is the remains of a once-living plant or animal whose hard parts (those resistant to decay, like shells, bones, claws, and teeth) have been preserved within the Earth. Clams and mammals, then, have a better fossil record than soft creatures like worms or slugs.

> A **fossil** is formed from the remains or traces of plants and animals preserved by natural causes in the Earth's crust.

The key to fossil preservation and dating is decomposition. If a living organism is buried quickly before oxygen and microorganisms can break it down, it has a good chance of being preserved. Since shells and bones are fairly porous, minerals from surrounding sediments are deposited in open areas where the once fleshy parts have dried or disintegrated. As we learned in Chapter 2, geologists find the age of fossils by radioactive dating. The better preserved the sample, the more accurate the dating within certain margins of error.

Skin imprints, tracks, and feces (called *coprolites*) from ancient organisms are also considered to be fossils. Can you think of any other profession where the rock-hardened waste from ancient animals is considered a real find?

By studying petrified stomach contents and coprolites (measuring from less than a centimeter to over 30 cm) from dinosaurs, paleontologists have gotten information about the diet of dinosaurs. Because some plants are tough and resistant to digestion, some coprolites are pretty intact. Several plant-eating dinosaurs had a wide diet by today's salad standards.

Tooth shape also gives clues as to diet. Scientists look at modern animals' tooth shape for eating habit and diet comparisons. Broad, grinding teeth seem to point to plant eaters, while jagged, sharp teeth are used by meat-eating predators. Figure 10-1 gives you a sampling of different dinosaurs that have been categorized into plant eaters, meat eaters, or omnivores (eats both plants and animals).

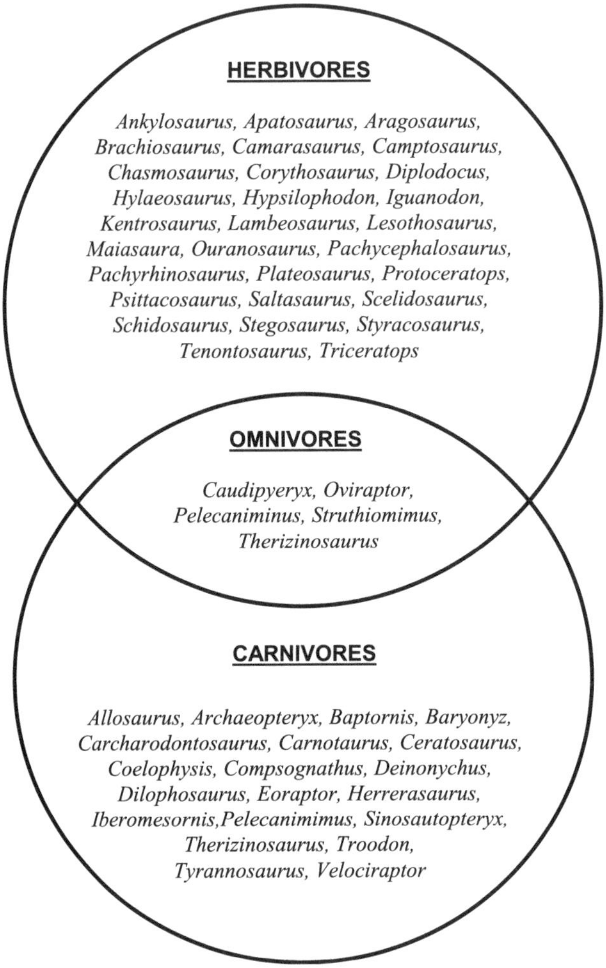

HERBIVORES

*Ankylosaurus, Apatosaurus, Aragosaurus,
Brachiosaurus, Camarasaurus, Camptosaurus,
Chasmosaurus, Corythosaurus, Diplodocus,
Hylaeosaurus, Hypsilophodon, Iguanodon,
Kentrosaurus, Lambeosaurus, Lesothosaurus,
Maiasaura, Ouranosaurus, Pachycephalosaurus,
Pachyrhinosaurus, Plateosaurus, Protoceratops,
Psittacosaurus, Saltasaurus, Scelidosaurus,
Schidosaurus, Stegosaurus, Styracosaurus,
Tenontosaurus, Triceratops*

OMNIVORES

*Caudipyeryx, Oviraptor,
Pelecaniminus, Struthiomimus,
Therizinosaurus*

CARNIVORES

*Allosaurus, Archaeopteryx, Baptornis, Baryonyz,
Carcharodontosaurus, Carnotaurus, Ceratosaurus,
Coelophysis, Compsognathus, Deinonychus,
Dilophosaurus, Eoraptor, Herrerasaurus,
Iberomesornis,Pelecanimimus, Sinosautopteryx,
Therizinosaurus, Troodon,
Tyrannosaurus, Velociraptor*

Fig. 10-1. More herbivores than carnivores are known and only a few omnivores.

Fossil Rock Types

Highly sensitive instruments like mass spectrometers are used to find the chemical makeup of fossil bones and coprolites. From these data, daily diet and mineral composition can be retrieved and applied to plants containing high levels of the same minerals.

Sedimentary rocks that contain fossils were laid down as sediments on floodplains by rivers, streams, lakes, and inland seas. Sediments are formed from layered sand, mud, and plant and animal materials. This preserves and protects prehistoric remains from air and deterioration. Over time, sediments harden and compress into sedimentary rock.

Since igneous rock is formed from super hot magma, fossils are not likely to be found in igneous rock. Metamorphic rock that was originally sedimentary rock contains fossils, but because of change and compaction, fossils are destroyed and distorted.

Fossil-containing rock can also be heated during metamorphism. This heating destroys the smaller, more delicate fossils first, followed by the heavier, sturdier ones. However, like other exceptions in nature, some fossils have survived high heat and pressure depending on specific circumstances.

Tectonic processes destroy fossils as well. Rock pressures stretch and distort fossils just as they do to other types of rock. Cleavage that develops perpendicular to parallel layers of sedimentary rock fractures fossils as well. Shearing, shifting, and uplifting of landmasses smash fossil-containing rock beyond recognition, leaving little behind for geologists to study.

When dating fossils and rock layers, the oldest layer will be at the bottom, while the newest layers are on top. This is true unless a sedimentary rock layer has been compressed or uplifted by tectonic forces. This oldest-to-youngest layering is called *superposition*.

It is possible for a part of a rock to be incorporated into another sedimentary rock layer that is older. This can be seen when younger rocks or other geologic features cut across older, preexisting rocks. When this happens, it is known as an *inclusion*.

Sometimes, geologists find that erosion has erased a large part of a rock layer. There are missing pieces. These are erosion or nondeposition surfaces that represent a certain time period and must be accounted for in other ways.

> Significant breaks in the geologic record due to erosion or a lack of sedimentary deposition are called **unconformities**.

The same thing can happen with fossils. Fossils at the bottom of a sedimentary rock stack are older than the layers above it. The gradient of change in fossil types is also oldest to youngest.

> **Fossil succession** is when older fossils are found in the lowest levels of sedimentary rock layers while more recent fossils are found closer to the surface.

Climatic Changes

Another factor in fossil scattering, besides tectonic movement, is the changing climatic cycles. We saw in Chapter 2 that over geological time, the level of the oceans has risen and fallen many times. The water overflows the land and then drains off again. While this is going on, the climate was also heating up at times and then cooling down again. The world and its citizens enjoyed hot and muggy conditions during some time periods and cold and freezing during others.

Some climatic changes were worldwide, while others affected some landmasses more than others, depending on location. If a continent originally located at the equator drifted northward over a geological time, then its resident plants and animals would have to change to survive.

Geologists believe that there have been four major periods of glacial activity during the time that most fossils were deposited and transformed into rock. The first of these took place in the Precambrian era and before. It is estimated that the Precambrian represents 88% of all of known geologic time. It is divided into the Hadean (beginning), Archean (ancient), and Proterozoic (early life) eons. The Phanerozoic eon (visible life) is the youngest eon with the best fossil and rock record. The majority of what paleontologists study today is from the Phanerozoic. The different eons, eras, periods, and epochs can be a bit confusing. To refresh your memory about the different geological times, turn back to Table 2-2 in Chapter 2.

The second glacial phase happened late in the Ordovician (Phanerozoic eon, Paleozoic era). The third important glacial age took place during the Carboniferous to middle Permian time, estimated at around 20 million years. During this deep freeze, the southern hemisphere, containing (bunched at that time) South America, Africa, India, and Australia, was one big polar ice cap.

During the melting of this ice chunk, huge amounts of water flooded existing primeval jungle and plants, amphibians and insects habitats. Fossil sedimentary rock layers contain these types of inhabitants along with the marine organisms that were carried in by the flood waters. The final glacial period began in the Pleistocene period (around two million years ago) and continues on today.

Thorough study and radiometric dating of fossil pollen, organisms, and surrounding rock have made finer glacial time divisions possible. Now geologists believe that there may have been greater than 15 periods of glacial ebb and flow. These periods included warmer climates with thriving, diverse species found throughout. However, geologists are finding that a richer line-up of species seem to be found during times of various temperature changes and glacial cold.

Species, alive at various periods in the Earth's history, had to adjust to climatic changes during their lifetimes. The strongest passed on their genetic information to their offspring who lived other lifetimes. Table 10-1 gives you an idea of the dinosaur record holders that have been discovered.

Time is the key to change. Think of it like surviving a volcanic eruption. The more time you have to get away, the better your chances of survival. This is one of the problems that species on the endangered list face today. Their habitats are changing (sometimes eliminated completely) within a few short years and mostly they don't have time to adapt or escape. That is why it is important if we are to preserve plant and animal species for future

Table 10-1   Paleontologists are always looking for the next unique fossil discovery.

Characteristic	Record holder
Shortest	*Compsognathus* (1 m)
Heaviest	*Argentinosaurus* (100 tons)
Tallest	*Brachiosaurus* (12 m)
Largest head	*Pentaceratops* (horned, skull over 3 m long)
Longest	*Seismosaurus* (50 m)
Largest meat eater	*Giganotosaurus* ($14\frac{1}{2}$ m long)

generations for people to take a hard look at how the development of natural areas affects native species and habitats in the area.

Mineralization

Have you ever found a piece of petrified wood? If you have, you can understand why early humans were puzzled. It looks like wood, with wood grain and often bark, but it's hard like a rock. In fact, it is rock! How can wood turn into rock? The answer is *mineralization*, also known as *petrification*. The cells and biological parts of the wood are replaced over time by minerals from surrounding rock.

> **Mineralization** or **petrification** happens when the organic parts of an organism are replaced by inorganic minerals.

This happens when the original bone or shell material (calcium phosphate) is replaced by minerals (silica) transported by ground water through cracks and crevices in the rock covering the specimen. Sometimes this process of fossilization takes place perfectly with nearly every detail preserved (like spines or tendrils on shellfish) and other times, just the large parts are preserved. It depends on the environment at the time the organism died and what happened to it over time.

Just as minerals are different colors (iron is red, chlorite is green), fossilized bones are often different colors depending on the type of minerals in the surrounding sedimentary rock.

Fossil Naming

The *International Code of Zoological Nomenclature* (*ICZN*) is the organization that determines the naming of an organism. It has specific rules, which maintain a standard of naming that all scientists know and understand.

All organisms, modern, ancient, and everything in between are classified into major groupings, according to specific characteristics. The more significant distinction is between the single-celled, no nucleus organisms (*Prokaryotes*) and multicelled, with a nucleus organism (*Eukaryotes*). These are the two major groups to remember with everything else placed below them.

Only one large category, *Monera*, is found below the Prokaryotes group. It contains bacteria and blue-green algae.

The Eukaryotes are divided into four major categories: the *Protista*, *Fungi*, *Plantae*, and *Animalia*. The Animalia category is further divided into *Vertebrata* (with a spinal column or backbone) and *Invertebrata* (no spinal column or backbone).

Paleontologists study fossils singly, but always within an ICZN-defined category. In fact, all living organisms are divided into separate groups and subgroups. When a plant or animal is named, for example, it is categorized within a wide system of narrower and narrower characteristics. When it can't be defined any more, you are at the most basic level and scientific name. The complete ICZN classification system has hundreds of branches describing in finer and finer detail the differences between species.

Scientific names for different organisms have two parts. The first name, the *genus*, is always capitalized, while the *species* name is written second in lower-case letters. Both genus and species names are always italicized like in *Centrosaurus apertus*. Additionally, the genus name can be abbreviated like in *A. fragilis* (*Allosaurus fragilis*) or used alone to describe many species in one genus (*Allosaurus*).

Homo sapiens are the genus and species names for humans, while *Tyrannosaurus rex* is the genus and species of a meat-eating dinosaur. In this study of fossils, we will only name fossils to their genus level. To give you an idea of the main ICZN categories used to name fossils and living organisms, see Fig. 10-2.

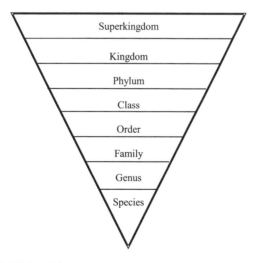

Fig. 10-2.   All living things are classified within a standard naming system.

Humans are categorized in the following way:

- Superkingdom → Eukaryota
- Kingdom → Animalia
- Phylum → Chordata
- Class → Mammalia
- Order → Primates
- Family → Hominidae
- Genus → Homo
- Species → sapiens

When paleontologists talk about other fossil specimens in the Hominidae family, they may use the genus name *Homo*, but with different species names like *Homo erectus* or *Homo habilis.*

Reconstructing Fossils

Figuring out how a dinosaur might have looked from a site of scattered bones can sometimes be tricky. It's like putting together a puzzle of a picture you have never seen and where some of the pieces are missing. From one discovery to the next, it sometimes takes decades to add another piece to the puzzle.

One famous goof happened when paleontologist, Gideon Mantell, was putting together the skeleton of the dinosaur, *Iguanodon*. He got it pretty much right, but put a small bone/horn at the end of the *Iguanodon*'s nose, that turned out later (by comparing to other newly uncovered *Iguanodons*) to actually be its thumb!

Reconstructing large plants is also tough. When plants die, they often come apart and fall in a heap. There is no good way to figure out what were branches, roots, or leaves, let alone where they connected together. The fact that different parts are found intact at different times leads paleontologists to name them as they are found. So the same plant may have different names for its flowers, leaves, and roots, until one day a complete specimen is found and all known information falls into place. In some locations, fossils are pretty much everywhere, and then in others it may be years, if ever, before another specimen of the same type are found again.

Fossil Finds

How do you find a fossil? Well, there are millions of fossils in the rock layers beneath the crust's thin blanket of soil. The best place to look is in an exposed area of sedimentary rock. Then, armed with a compass, geological map, magnifying glass, ruler, geological hammer, toothbrush, large paintbrush, chisel, trowel, plastic bags, boxes, packing paper, tape, notebook, marking pen, back pack, a field guide to fossils and a healthy appreciation of the fragility and importance of fossils, an amateur can begin the quest for prehistoric shark's teeth, ammonites, trilobites, or whatever comes to light. Fossils have often been found by farmers and construction crews who unearthed them by accident. Just think of the treasures of nature accessible to the properly prepared geologist.

One important thing to remember is that a special permit is required to remove fossils from public land. So, besides the necessary geological tools, make sure you have a permit or ask permission from landowners before stomping all over their property and removing fossils. If you do, you will probably be invited back!

It is also possible to attend paleontological digs and assist scientists with the time-consuming and tedious job of carefully removing, mapping, and preparing fossils from the best known and most significant fossil sites around the world.

Como Bluff in southern Wyoming, near Medicine Bow, is famous for its Jurassic dinosaur fossils. First discovered in the 1870s, by Union Pacific workers building the transcontinental railroad near the base of the bluff, the area quickly sparked the interest of paleontologist, O.C. Marsh of Yale University in 1877, who led expeditions over the next 10 years unearthing hundreds of dinosaur skeletons.

In 1898, Walter Granger, from the American Museum of Natural History began working with a team near Como Bluff. The place was called Bone Cabin, because there were so many bones that a local sheepherder built a cabin nearby using fossils as building material!

A few other dinosaur bone and fossil sites where many excellent remains have been found, including entire nests of eggs, are listed in Table 10-2.

In 1999, the jaws from two of the oldest dinosaurs ever discovered, and the remains of eight other prehistoric animals, were discovered in an area rich with fossils in Madagascar. The bones were thought to be from the mid-late Triassic Period. The site included the fragmentary remains of two plant-eating dinosaurs, *prosauropods*, about the size of a large dog.

Table 10-2   Dinosaur bones have been found in abundance all over the world.

Geographical location	Geological time period	Specimens	No. of specimens found
Como Bluff (Wyoming, USA)	Jurassic	*Diplodocus*	100s–1000s
Bone Cabin Quarry (Wyoming, USA)	Jurassic	*Apatosaurus, Ornitholestes, Camptosaurus, and Stegosaurus*	100s–1000s
Hell Creek (Montana, USA)	Late Cretaceous	*Tyrannosaurus, Albertosaurus, Triceratops, Anatotitan, Thescelosaurus, Pachycephalosaurus, and Edmontosaurus*	100s
Red Deer River (Alberta, Canada)	Late Cretaceous	*Corythosaurus, Apatosaurus, Saurolophus, Struthiomimus, Albertosaurus, and Dromaeosaurus*	100s
Burgess Shale Quarry (British Columbia, Canada)	Proterozoic and Cambrian	*Yuknessia* (green algae), *Rhodophyta* (red algae), *Ogygopsis, Olenoides* (trilobites), *Marella* (lace crab), *Waptia, Naraoia, Vauxia* (sponges), *Canadia sparsa* (annelid), *Hallucigenia* (velvet worms) (over 170 marine species identified)	100,000s
The Flaming Cliffs (Mongolia)	Cretaceous	*Protoceratops, Pinacosaurus, Saurornithoides, Velociraptor, Oviraptor*	<100
Western Gobi (Mongolia)	Cretaceous	*Protoceratops, Oviraptor*	<100

The fossilized bones are thought to be the earliest dinosaurs found in Madagascar and perhaps the earliest dinosaurs found anywhere in the world.

The prosauropods, plant eaters with long necks and small heads, could walk on two or four legs. These early dinosaurs either shared a common ancestor with, or were ancestors to, the huge sauropod dinosaurs that came later, such as *Apatosaurus*.

The Middle to Late Triassic (225–230 million years ago) is a time that has a sketchy fossil record. Paleontologists have found that at the start of this period, various reptiles, amphibians, and other vertebrates lived on the continents. By the end, early dinosaurs and mammals had appeared. The problem, until the Madagascar find, was that the skimpy fossil record left paleontologists with few clues about development and species during the in-between years.

Paleontologists from the Madagascar expedition also hope to discover clues to the breakup of the supercontinent Pangea, which seems to have begun in the Triassic period.

Fossils Found Together

If paleontologists discover a sedimentary rock cliff full of a variety of fossils, what can they predict about their interaction?

> A **fossil community** is a group of fossils of organisms that lived in the same environment when they were alive.

In a particular area, rocks formed very specifically. Their specific characteristics are called *sedimentary facies*. This happens today as it has for millions of years. Sedimentary rock in different places can have completely different compositions even though they may have accumulated during the same time period. The same thing happens with their associated fossils.

The fossils in some sites are often very different from fossils at another site because of the environment in which the sediment was deposited. Just as ocean fish, shellfish, and mammals are found at different sea depths, the fossils discovered from sedimentary deposits of former seas are characteristic of their marine environments. The sedimentary rock record often shows a species and sediment gradient. Since everything eventually sinks to the bottom and is covered by sediment, even species that lived far apart (in depth) may end up close together in the sedimentary rock layer.

Different sedimentary facies may have different groups of fossils. When different fossils are found together in one particular sediment type, those species are known as the *facies fauna* for that sediment.

Fossils in an *assemblage* can be from more than one fossil community. Individuals may have lived at the same time, but are located in broader strata that can range over millions of years.

A **fossil assemblage** is a group of naturally associated fossils found in an area of sedimentary rock strata.

Depending on sediment thickness, texture, color, orientation, and other factors, paleontologists can figure out if different individuals of the same genus and species lived at the same time through radiometric dating. They can also compare individuals from different classifications (kingdom, phylum, family, or whatever).

Paleontologists use fossils to date the sedimentary rock layers in which they are found. They look for similarities between fossil specimens in different layers and then apply this information to their knowledge of the climate and history of the region. This is how they were able to figure out which dinosaurs lived during the Jurassic, Cretaceous, and other periods.

A **correlation** happens when two or more rock units are dated on the basis of having similar types of fossils. Those with like fossil assemblages are thought to have formed around the same time.

Paleontologists have found that dinosaurs were not alone on the Earth. During the Mesozoic era, they lived with a wide variety of other animals. These animals were a lot like groups of animals alive today. The first mammals appeared in the Late Triassic along with frogs, turtles, lizards, and crocodiles. By the end of the Cretaceous period, birds, snakes, and primitive marsupials and mammals were around. They all lived together in a region sharing water holes and food (and being food) like birds and animals, large and small, do today.

One problem paleontologists face when creating a total picture of a fossil assembly is that many marine tube worms and other species are soft bodied. Marine plankton, tube worms, and the like have no hard bones. Their remains decompose quickly. The only clue to their existence is the fossilized tunnels left behind by burrowing creatures. So some species have to be assumed by their remaining habitats.

Fossil Tracks

In 1802, a Massachusetts farm boy named Pliny Moody found fossil foot-prints in his family's field while plowing. Local farmers nicknamed the tracks, the footprints of "Noah's Raven." In 1835, James Deane, a Connecticut doctor and amateur naturalist, saw two side-by-side sets of footprints in rock slabs cut for a roadbed near his home. He thought they had been made by turkeys. Deane was wrong about the maker of the tracks, but was probably one of the first people to suggest that living animals with very similar tracks to the fossil imprints might be very similar to extinct creatures in other ways.

From the Paluxy River region in Texas, tracks were found by R.T. Bird. In 1937, when Bird visited the area, he found that fossil tracks were common on local ranchland. Among these were the Davenport Ranch Trackway and Paluxy River trackways.

At the Paluxy Trackway, Bird discovered the tracks of more than 12 sauropods (*Brachiosaurus*), followed sometime later by at least three predators (theropods like *Allosaurus*). Although Bird thought this was evidence of a hunt in progress, there is no way to tell how much later the predators followed. Like most dinosaur detective work, it's impossible to know. The track could have been a favorite sauropod path that the theropods took as a shortcut.

Dating Rock and Fossils

As we learned in Chapter 2, geologic time is calculated by measuring the decay rates of the atoms of radioactive isotopes. The ages of the most ancient rocks can be found by measuring the decay of specific isotopes that are not stable, but break down to other element forms. A rock sample is dated using testing techniques known as radiometric dating, which takes into account the various melting and environmental influences that affected the sample.

Unstable isotopes of elements like uranium, rubidium, potassium, and carbon go through natural decay into stable isotopes of other elements. This is the case when atoms of an unstable isotope of uranium (parent isotope) decay into atoms of a stable isotope of lead (daughter isotope). The ratio of the uranium isotope to stable lead isotope in a sample is a function of time. A longer time period results in a higher percentage of daughter isotopes compared to the smaller percentage of parent isotopes. The decay rates of

radioactive breakdown products are constant and can be accurately determined by the laboratory testing of a sample.

The discovery of radioactivity in 1896, by Antoine Becquerel, led to the study of radioactive decay of elements in fossils. By measuring decay rates of various unstable isotopes geologists were able to add numbers to the traditional geologic timescale. Today's measurement techniques and extremely accurate instrumentation have allowed for precise fossil and sediment dating. The geological timescale is constantly updated with new finds producing time adjustments and increased accuracy.

Quiz

1. A group of fossils that lived in the same environment when they were alive are called a fossil
   (a) club
   (b) community
   (c) habitat
   (d) find

2. *Homo sapiens* are the
   (a) names of rock music groups
   (b) kingdom and phylum names for humans
   (c) genus and species names for humans
   (d) genus and species names for whales

3. Paleontologists think the breakup of the supercontinent Pangea began during which period?
   (a) Cambrian
   (b) Mesozoic
   (c) Jurassic
   (d) Triassic

4. Fossils in an assemblage
   (a) are easy to find
   (b) can be from more than one fossil community
   (c) are never seen
   (d) can be from only one fossil community

5. A fossil is formed from
   (a) children's clay
   (b) the remains of plants and animals preserved by natural causes in the Earth's crust

(c) igneous rock

(d) seawater that has dried and formed crystals

6. When the organic parts of an organism are replaced by inorganic minerals, it is called
    (a) ionization
    (b) mass wasting
    (c) mineralization
    (d) organization

7. *Prosauropods* were what type of dinosaur?
    (a) plant eaters
    (b) meat eaters
    (c) snakes
    (d) hard shelled

8. Different fossils found together in one sediment type are known as the
    (a) Brontosaurus band
    (b) ammonites
    (c) facies fauna
    (d) differentiated flora

9. When older fossils are found in the deepest levels of sedimentary rock and younger fossils are found near to the surface, it is called
    (a) fossil succession
    (b) a lucky find
    (c) lithification
    (d) careful digging and a sharp eye

10. Sometimes, the only clue to the existence of ancient soft-bodied marine species is
    (a) a large coprolite sample
    (b) a forwarding address
    (c) the color of the sedimentary rock
    (d) the fossilized tunnels left behind

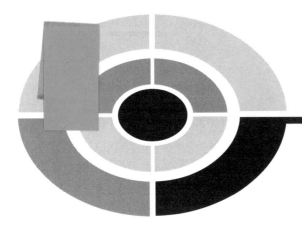

Part Two Test

1. Sudden dense movements of water that slice deep canyons along the ocean floor are called
   (a) whale tail wake
   (b) turbidity currents
   (c) geological uniformity
   (d) rapidity currents

2. The magnetic field strength or field intensity is the
   (a) force applied to a magnetic pole at any point
   (b) force of magma rising to the surface
   (c) force of the Jedi warrior
   (d) centripetal force of a spinning top

3. The chameleon type of rock is
   (a) igneous rock
   (b) protoplastic rock
   (c) sedimentary rock
   (d) metamorphic rock

4. The word metamorphic means to
   (a) change form
   (b) divide in half
   (c) cook thoroughly
   (d) stay the same

5. What is the main effect of weathering?
   (a) the change of seasons
   (b) wears away rocks
   (c) wrinkles
   (d) snow

6. Which of the following minerals is not commonly found in igneous rock?
   (a) silica
   (b) iron
   (c) beryllium
   (d) sodium

7. High-grade metamorphic rock is a result of
   (a) high temperature and low pressure
   (b) high temperature and high pressure
   (c) low temperature and low pressure
   (d) low temperature and high pressure

8. Shock metamorphism happens
   (a) over millions of years
   (b) in six months
   (c) in an instant
   (d) never happens

9. Metamorphism takes place in all of these places, except
   (a) surface
   (b) mantle
   (c) atmosphere
   (d) oceanic crust

10. A thermal gradient is defined as the
    (a) heat measured from hot concrete
    (b) amount of heat for magma to rise
    (c) temperature needed to keep ice cream frozen
    (d) rate of temperature increase compared to depth

11. Marble, quartzite, and hornfels are all examples of
    (a) soft rock
    (b) foliated rock
    (c) rock and roll
    (d) nonfoliated rock

12. When the organic parts of an organism are replaced by inorganic minerals, it is called
    (a) crystallization
    (b) petrification
    (c) organization
    (d) sedimentation

13. Which of the following is not a fossil?
    (a) arrowhead
    (b) trilobite
    (c) Tyrannosaurus rex
    (d) Triceratops

14. In sedimentary rock from ancient oceans, fossilized tunnels are left behind by
    (a) soft-bodied worms
    (b) buff, hard-bodied worms
    (c) ancient miners
    (d) tunnels are never found

15. The name *felsic* comes from a combination of the words
    (a) felt and sing
    (b) feldspar and sulfur
    (c) feldspar and silica
    (d) field ore and silica

16. Abyssal plain is an example of what kind of sedimentary environment?
    (a) marine
    (b) fissure
    (c) science fiction
    (d) mountainous

17. Igneous rock makes up what percentage of the top 10 miles of the Earth's crust?
    (a) 35%
    (b) 55%

(c) 75%

(d) 95%

18. Which of the following is not a way that metamorphic rock forms?
    (a) regional
    (b) hydrothermal
    (c) solar transformation
    (d) contact

19. An amateur paleontologist can often find
    (a) holes in their socks from all the walking
    (b) ancient shark's teeth, ammonites, and trilobites
    (c) tube worm fossils
    (d) the fossil bones of brachiosaurus

20. James Hutton thought that
    (a) the Earth was flat
    (b) hot magma made its way upward between layers of sedimentary rock
    (c) water explained the changes between sedimentary rock layers
    (d) he should have been a doctor like his mother wanted

21. According to the International Code of Zoological Nomenclature, humans are categorized in the kingdom
    (a) prokaryota
    (b) animalia
    (c) primates
    (d) sapiens

22. What is the only difference between a geode and a nodule?
    (a) a geode is solid inside and a nodule is hollow
    (b) geodes only contain amethyst crystals
    (c) a geode is hollow inside and a nodule is solid
    (d) geodes are found inside meteorites

23. When the first known *Iguanodon* fossil skeleton was assembled, a huge claw was mistakenly put on its
    (a) tail
    (b) neck
    (c) heel
    (d) nose

24. A mineral that light shines only partially through and is cloudy is said to be
    (a) translucent
    (b) transparent
    (c) opaque
    (d) priceless

25. Mafic rock
    (a) is seldom found on Earth
    (b) contains high levels of magnesium and ferric minerals
    (c) contains the same minerals as felsic rock
    (d) contains high levels of manganese and sulfur

26. The cycle of heat and pressure that transforms existing rock is called
    (a) thermogradient cycle
    (b) solar cycle
    (c) rock cycle
    (d) unicycle

27. Polarity reversal happens when
    (a) the core's magnetic field flows in the same direction as the layers above it
    (b) socks stick together coming out of the dryer
    (c) polar bears are really irritated
    (d) the North Pole's location and the South Pole's location switch places

28. A powdery stripe made when rubbing a sample across an unglazed surface is called a
    (a) foundation
    (b) streak
    (c) flute
    (d) habit

29. Talc, gypsum, calcite, fluorite, apatite, orthoclase, quartz, topaz, corundum, and diamond are standards on the
    (a) Alvandi scale
    (b) Fleming scale
    (c) Mohs' scale
    (d) Gregg scale

30. A paleontologist is a scientist that studies
    (a) light waves
    (b) paint

(c) horses

(d) fossils

31. Allochromatic minerals are generally

    (a) opaque

    (b) colorless and transparent

    (c) colorful and translucent

    (d) of a hardness of 1

32. Perfect, distinct, indistinct, or none are types of

    (a) cleavage

    (b) habit

    (c) luster

    (d) transparency

33. Andesite is named after the volcanic

    (a) Antarctic mountains

    (b) Rocky mountains

    (c) Himalayan mountains

    (d) Andes mountains

34. When a plutonic magma pocket is rimmed by a contact ring of metamorphic rock, it is known as

    (a) a volcanic vent

    (b) an aureole

    (c) a slipstream

    (d) breccia

35. Malachite is commonly

    (a) white

    (b) pink

    (c) deep blue

    (d) dark green

36. Dull and waxy are two different types of

    (a) luster

    (b) habit

    (c) geodes

    (d) specific gravity

37. When the crystals push against each other and form a mass, it is called

    (a) cleavage

    (b) massive

(c) contact twinning

(d) rhombohedral formation

38. Cinnabar, molybdenite, pyrite, and galena are minerals in which mineral group?

(a) phosphate

(b) sulfide

(c) silicate

(d) carbonate

39. What is one of the best known and most frequently seen intrusive igneous rock types?

(a) granite

(b) sand

(c) sediment

(d) tundra

40. When igneous intrusion of magma heats up surrounding rock, it is called

(a) lithification

(b) sedimentary rock

(c) contact metamorphism

(d) index mineralization

PART THREE

Surface News

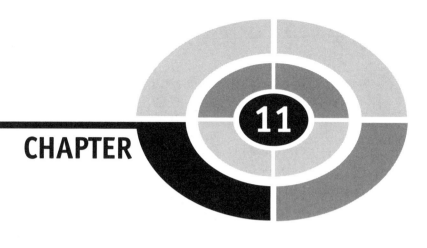

CHAPTER 11

Volcanoes

Suddenly the ground jerks and a roar like a thousand jet engines is heard. The air becomes a searing furnace with smoke and ash raining down from the sky. Is this something from a science fiction novel or "movie magic?" No, it's just another of Mother Nature's temper tantrums, known commonly as a *volcanic eruption*.

Volcanoes have erupted with hot magma, from pressure buildup in the Earth's interior, since the planet was formed. Magma from deep within the earth is sent out onto the surface in many different forms and intensities.

> A **volcano** is a mound, hill, or mountain formed from hot magma exiting the Earth's crust and piling up on the land or beneath the seas.

Ancient humans thought volcanic eruptions were caused by their bad behavior and the gods' resulting displeasure. Some cultures even thought that if they threw some unlucky sacrifice into the crater, that the gods would get happy again. Unfortunately for the sacrifice and any villages in the path of an erupting volcano, this tactic didn't work very well.

Native Americans in the Pacific Northwest told early explorers a legend about the active Mount St. Helens. In fact, the Indian name for the mountain, Louwala-Clough, means "smoking mountain." According to one legend, the snow-capped mountain was once a beautiful maiden. When two sons of the Great Spirit fell in love with her, she couldn't choose which one she liked best. The two braves fought over her, burying villages and forests in the process. The Great Spirit was not pleased. He smote the three and erected a mighty mountain peak where each fell. Because the maiden was beautiful, her mountain (Mount St. Helens) was a beautiful, symmetrical cone of dazzling white. One brave (Mount Hood) lifts his head in pride, but the other (Mount Adams) wept to see the beautiful maiden wrapped in snow, so he looks downward.

In the days of the Romans, Vulcan was not the home planet of a race of pointy-eared aliens of television's and movies' "Star Trek," but instead the name of the Roman god of fire and the blacksmith to the gods. Long ago, it was thought that an ancient island off the coast of Sicily was the location of Vulcan's home and smithy. Local people thought whenever neighboring volcanoes smoked, Vulcan was heating his forge. When noise and vibrations were felt in nearby villages, people thought Vulcan was hard at work hammering on his anvil. The only problem was that sometimes he got carried away and caused an eruption. The word *volcano* comes from the Latin word *vulcanus* which means "fire breathing."

The Romans experienced this in a big way when the city of Pompeii was destroyed by the eruption of Mount Vesuvius in AD 79. The inhabitants are thought to have had notice of the explosion, but chose to stay in the city, only to be buried by the unbreathable, searing ash.

When referring to volcanic materials ejected during an eruption, most *volcanologists* talk about *volcanic dust*, *ash*, *cinders*, *lapilli*, *scoria*, *pumice*, *bombs*, and *blocks*. These will be explained separately as different eruption types and volcanoes are described. If you want to talk about all the things that shoot out of a volcano, the word *tephra* is used. Table 11-1 shows the size differences of single and combined tephra.

> **Tephra** includes all the different types of matter that is sent blasting out of a volcano compared to bubbling, flowing **lava**.

In 1751, French geologist, Jean Guettard discovered evidence of ancient volcanic activity in the south of France near Auvergne. The highest peak in that chain of 50 extinct volcanoes is Puy de Dome at 1465 m. No one thought volcanoes had ever existed in Western Europe and scientists were stunned at

Table 11-1 Tephra rocks and tephra clumps have different names depending on size.

Particle size (mm diameter)	Tephra (single)	Pyroclastic rock (combined material)
< 2 mm	ash	ash tuff
2–64 mm	lapilli	lapilli tuff
> 64 mm	bombs	agglomerate

the time of Guettard's news. Ten years later, Nicholas Desmarest became so interested in Guettard's discovery that he began mapping newly recognized lava flows and volcanic mounds in the area. Since they had been worn mostly away by erosion, people had not previously realized what they were.

Mount St. Helens

In March of 1980, Mount St. Helens was considered one of the most beautiful mountains in the Pacific Northwest of the United States. Campers, hikers, and fishermen didn't think about the fact that the striking mountain in central Oregon sits on a plate boundary and had erupted as recently as 400 years earlier. Then, after a series of over 170 increasingly stronger earthquakes warned scientists of the danger, the top of the mountain was blown off in a tremendous explosion sending a column of volcanic ash and steam into the sky.

> **Volcanic ash** ejected during a volcanic eruption is made up of rock particles less than 4 millimeters in diameter. Coarse ash is sized from $\frac{1}{4}$–4 mm, with fine ash (dust) measuring $< \frac{1}{4}$ mm in grain size.

Two months later, after nearly constant tremors, a large bulge on the mountain's side appeared. Then, on the morning of May 18, 1980, an earthquake a mile underground lowered the internal magma pressure within the volcano and caused the bulge to collapse, followed by a huge landslide moving at 241 kilometers per hour. The removal of the overlying weight allowed the pressurized magma below to cut loose with a violence calculated at over 500 times the force of the Hiroshima bomb or 10 million tons of TNT.

The blast from the volcanic vent sent ash into the air to an estimated height of 19 km. The amount of lumber flattened in the 373 km area affected by the blast could build more than 250,000 homes.

Since Mount St. Helens' eruption happened early on a Sunday morning, many of the visitors and lumber people usually near the mountain had not yet arrived. Sixty people, including a number of volcanologists, died from the blast and hot choking ash. Of the local animal population, 5000 black-tailed deer, 200 black bear, 1500 elk died in the disaster, along with nearly all the birds and small mammals within the blast area.

Spokane, Washington, over 320 kilometers away, was in total darkness by mid-afternoon from falling ash. The wind-driven ash traveling at nearly 100 km/h caused power failures and clogged automobile and emergency vehicle air filters.

It is estimated that over 540 million tons of ash was spewed from Mount St. Helens over a 35,410 km^2 area, with most of the ash dropped on Oregon, Washington, and Idaho. In the out-lying cities hardest hit, as much as three inches of ash coated the countryside.

Mount St. Helens, along with Mt. Rainier, Mt. Hood, and others are part of the Cascade mountain range, formed where the western edge of the North American plate overrides the descending San Juan de Fuca plate. This fairly young mountain range, which lies along the eastern edge of the Pacific Ring of Fire, began forming between 3 and 7 million years ago when magma began rising through conduits to the surface. Table 11-2 lists the volcanic peaks that lie along the Cascade mountain range.

Lahars and Surges

Eruptions of snow-capped volcanic mountains with their super-heated ash melt peak snow and ice causing a mudflow of melt water and volcanic ash called a *lahar*. Lahars are extremely dangerous as they travel and kill quickly, burying everything in their path. Even when minor lahars flow downhill and clog existing streams, they cause a lot of flooding down-stream.

The worse disaster caused by a lahar happened in 1985 when the Columbian volcano, Nevado del Ruiz erupted. The melting and cracking of its summit snow and ice caused a 40-m lahar traveling around 25 km/hr to surge over the town of Armero. Approximately 23,000 people were killed instantly. The mass and force of the lahar leveled buildings, vehicles, and trees in its way. The violent mudflow covered the town to a depth between 2 and

Table 11-2   The Cascade Mountain range extends across 3 states and Canada.

Mountain or crater	Elevation (meters)	Location
Mt. Garibaldi	2678	British Columbia
Mt. Baker	3285	Washington
Glacier Peak	3213	Washington
Mt. Rainier	4394	Washington
Mt. St. Helens	2950 (before eruption) 2549 (after)	Washington
Mt. Adams	3742	Washington
Crater Lake (Mt. Mazama)	4394	Oregon
Mt. Hood	3426	Oregon
Mt. Jefferson	3199	Oregon
The Three Sisters (Paulina Pk.)	3157	Oregon
Sacajawea Peak	2999	Oregon
Mt. Moloughlin	2894	Oregon
Mt. Thielsen	2799	Oregon
Newberry Crater	2434	Oregon
Mt. Shasta	4401	California
Lassen Peak	3187	California

5 meters thick. The same thing happened along the same path in 1595 and 1845. Not a lucky place to live!

Although Mount St. Helens' volcanic blast was amazing, there have been larger blasts in recent history. For example, the 1815 eruption of Tambora in Indonesia ejected around $30\,km^3$ of debris – roughly 30 times the volume of the *pyroclastics* of Mount St. Helens.

Pyroclastic surges are the most violent of all volcanic activities. A pyroclastic surge can travel with the speed of an atomic blast in the form of low-density, extremely hot clouds of rock particles and gas flows. They travel at over 161 km (100 mi) per hour and create huge turbulences.

In 1902, Mont Pelée erupted near the town of St. Pierre, Martinique located 8 km away. It sent a fast-moving, veil-like cloud of hot volcanic ash and gases over the town and leveled it. The cloud destroyed all plant life, wiped out the town and killed 30,000 people.

> **Pyroclasts** are fragments of crystallized volcanic rock and previously solid lava of any size that are forcibly ejected from a volcanic vent during an eruption.

A *pyroclastic flow* levels and chars everything in its path including trees and buildings. Like the Mount St. Helens eruption, an eruption's pyroclastic flow can be more deadly than traveling lava. These flows are of higher density and follow the contours of the land more closely than surges. Their combined heat and speed make them lethal. It is nearly impossible to get out of the way of a pyroclastic flow. The few people, who have survived such eruptions, were able to find shelter behind a ridge or other natural feature and were very lucky to survive the heat.

> A **pyroclastic flow** is the denser-than-air mixture of fine ash and hot gases (temperatures over 1000°C) resulting from an eruption.

Pyroclastic Matter

A *pumice fall* is made up of bubbling magma blown out of a volcanic vent or fissure by superheated gases. The change in pressure and expansion of gases in the magma, when it reaches the surface, creates cavities throughout the rock. This magma-type hardens into *pumice*, a light volcanic glass containing many small air bubbles that allow it to be carried great distances on the volcanic wind.

When pyroclastic fragments cement together to form rock, they are named according to the size of the fragments. When particles are small (less than 2 mm across) they are called *ash* and the rock is called an *ash tuff*. Particles that are bigger, between 2 and 64 mm in diameter, are called *lapilli* and the rock formed is called *lapilli tuff*. The largest pyroclastic particles, larger than 64 mm across, are called *bombs* and the rock formed is called an *agglomerate*.

The particles of ash in some volcanic eruptions are very hot. When they land, the extreme heat is so hot that it melts particles together. This glassy, fine-grained rock formed by the fusion of settling volcanic ash is called *welded tuff*. Some welded tuff natural glasses look a lot like lava flows. Volcanic rocks in the western United States that look like cooled lava are really welded tuffs. Some tuffs and agglomerates are layered and look a lot like sedimentary rock. When more than one layer of ash is spread by wind and rain, it may become solid in individual layered volcanic beds. If loose volcanic fragments are washed down the steep slopes of a mountain as mudflows and debris flows, the settling rock might be partly igneous and partly sedimentary.

Another type of pyroclastic flow makes a type of glassy, melted ash called *ignimbrite*. This type of flow is very large and covers hundreds of square kilometers of land to depths of several feet. When ancient Mount Vesuvius erupted in August of AD 79, it covered the towns of Pompeii and Herculaneum with chunks of hardened lava and thousands of pounds of ignimbrite ash. It did not cover it by a lava flow. Vesuvius has been active off and on over the past 300 years with the last eruption coming as recently as 1906. The large Italian population keeps returning and rebuilding in the area, even though the mountain could erupt again.

As we learned in Chapter 6, igneous rock is formed from the cooling and hardening of magma within the earth's crust. Over 95% of the top 16 km of the crust is made up of igneous rock formed from lava eruptions. That is a lot of volcanic activity! Let's see how all that happens.

Lava

When molten rock (magma) rises to the surface and flows out onto the land or into the oceans, it is called *lava*. After lava cools and hardens, it becomes rock. Figure 11-1 shows the path that magma takes to the surface to become a volcano. By forcing its way out of the magma chamber deep within the earth, it makes its way upward. Heat and increasing pressure keep pushing and melting the rock in the way, until finally magma erupts at the surface as lava or tephra or under the sea in a boiling gush.

When Mount St. Helens in Oregon, Mount Vesuvius in Italy, and Mount Pinatubo in the Philippines erupted at different times in history, human kind realized how powerless it was when compared to the tremendous forces of nature.

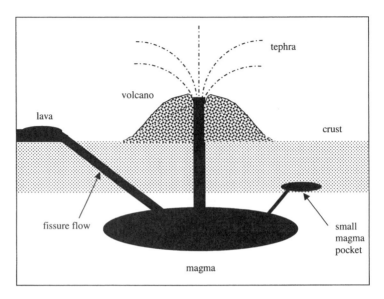

Fig. 11-1.   Magma surges to the surface in a volcanic eruption.

BASALTIC LAVA

Basalt is the most common volcanic igneous rock in the Solar System. The majority of the crusts of Mercury, Venus, Earth, the Moon, and Mars are made up of basalt.

Basaltic rock is an extrusive rock found over much of the earth's crust. It is composed mostly of *flood, pillow, pahoehoe,* and *Aa* lava. The different types and textures are formed according to how fluid the different lava types are during an eruption and how quickly they cool to form rock.

Nearly all of the Earth's ocean floor is made up of basalt. It is a mafic rock with only about 50% silica by weight.

FLOOD LAVA

Flood lava is pretty much what it sounds like, a flood of molten rock. It is made up of running basaltic magma that erupts on flat land and spreads out in thin sheets as a flood of lava. When there are a bunch of different lava floods over a period of time, they pile up on top of each other forming thick basaltic lava plateaus, called *flood basalts*. The Columbia River plateau of Oregon, Idaho, and Washington in the United States is a huge flood basalt covering millions of square kilometers. The Deccan and Siberian Traps (flood basalts) of India and Asia are even larger.

PAHOEHOE LAVA

Pahoehoe and *Aa lava* types are made up of cooling basaltic lava that has flowed downhill. Pahoehoe is the Hawaiian word for "ropy" and it looks like long, thick, stringy strands of rope. It forms when fluid lava flows into sheets and forms a cooled glassy skin. As the molten lava beneath the surface moves along, it pulls and twists the surface skin into thin, sinewy, coiled folds.

AA LAVA

Aa lava is very different in form from pahoehoe lava. It looks like freshly dug earth, after it has been turned over in preparation for spring planting. However, it is really hot lava! It is said that someone trying to cross Aa lava in bare feet yells, "Aa! Aa!" when discovering the mistake. Aa lava has lost most of its gases and flows much more slowly and steadily than pahoehoe lava. A thick skin forms at the cooler surface and as the lava keeps moving, it cracks and breaks into jagged chunks and blocks. The moving flow piles up boulders in front of it like a bulldozer.

PILLOW LAVA

Pillow lava is formed from slow, gurgling underwater eruptions. These oval, blobs of lava look a bit like rounded sand bags of 20–100 cm in width. When geologists find pillow lava on land, they know that the area was once underwater. Divers have seen pillow lava formed on the ocean floor off the coast of Hawaii. Flows of molten basaltic lava form tough skins when they hit cold ocean water, but the lava on the inside is a different story. It cools much more slowly forming a crystalline structure, while the outside skin hardens to a smooth (non-crystalline) glass. As the pillow lava continues to flow, it pushes the hot inner lava through the cooled glass shell and cracks it. When this happens, a new oval, lava lobe is formed. As the flow continues, these lobes break off and cool as individual masses (pillows).

RHYOLITIC LAVA

This type of lava is the most felsic of the different types. It is a light colored lava and erupts at temperatures of 800–1000°C. It has a higher silica content and is much thicker than basaltic lavas. Think of the flow differences between milk and molasses. *Rhyolitic lava* moves about ten times slower than basalt and tends to pile up in thick globular deposits.

ANDESITIC LAVA

This type of lava is a middle-of-the-road lava between basaltic and rhyolitic. It has a median silica content and its characteristics fall in between basalts and rhyolites.

In 1943, the postmaster of the island of Hokkaido, Japan and an amateur geologist noticed some unusual geological activity in a nearby potato field. What first began as escaping steam, and then turned into spewing ash and lava became the beginnings of a new volcano. The postmaster, a geologist's assistant at one time, watched the volcano's progress through his office window. (That was before color TV and DVDs.) He noticed that the window screen surrounded his view like a piece of graph paper, so he made notes of the volcano's growth month after month based on the squares of the screen. When geologists later studied his notes, they liked the graphing idea and found it to be a great way to measure a volcano's progress. The postmaster's volcano, *Showa Shinsan* (Japanese for "new volcano") was formed over an active subduction zone along the Pacific Rim. It produced *andesite* lava. Andesite is a basalt containing a high silica content.

LAVA LAKES

Lava lakes are often a product of Hawaiian eruptions where fluid basalt will pond in vents, craters, and wide low spots. Sometimes lava pours from a vent within a crater or wide depression and fills it. Today, active lava lakes are found in only a few places: Mount Erebus in Antarctica, Erta Ale in Ethiopia, and Nyiragongo in the Congo. Hawaii's Kilauea volcano has had an active past with lava lakes in several of its craters at different times.

As a lava lake cools, a silvery crust of only a few centimeters thick forms on the surface of the lake. This crust is constantly disturbed and reformed. The movement of the molten lava beneath the crust causes it to crack into slabs that sink. The newly exposed lava cools to form another crust and the process begins all over again.

Vents

From the deepest levels of the mantle, cracks or openings to the surface from the magma chamber are called *volcanic vents*. Magma is forced upward through these paths, then ejected as lava or pyroclastic ejecta. A vent located at the top of a volcano and at the end of a vertical *magma feeder pipe* is known as a *central vent*.

Vents mark the beginnings of a volcano and can have different shapes, but are mostly rounded. When a vent opens as a long slit in the ground, it is called a *fissure*. Sometimes lava is ejected forcibly from these fissures, but they can also ooze lava out regularly as well.

Fire fountains are the amazing jet-like sprays of liquid lava that spurt from central vent Hawaiian type eruptions. These incandescent jets spray hundreds of meters into the air. They shoot in brief spurts or sometimes provide fantastic fiery shows for hours at a time.

In 1959, the greatest fire fountain ever recorded in Hawaiian history shot lava 580 meters into the air from the Kilauea Iki vent. Then in 1986, this record was nearly tripled when an eruption on the Japanese Island of Oshima sent a lava fountain 1600 meters into the air. When fire fountain sprays are blown away downwind, they often produce an airborne curtain of glowing fragments that showers downward in a fiery breeze.

When small pyroclasts are carried downwind in a fire fountain, they cool quickly into black, volcanic glass beads, teardrop shapes, and oblong forms, known as *Pele's tears*. When the wind is really blowing, the beads are sometimes pulled into long glass fibers. Then, when the fibers are broken off from the main bead, they are called *Pele's hair*. I wonder if these glass fibers gave engineers the idea for fiber optics.

Sometimes after all the magma is used up, volcanic vents keep releasing gas fumes and steam through small vents called *fumaroles*.

Circulating groundwater that seeps down to hot magma is super heated and sent back up to the surface as hot springs and *geysers*. A geyser is a hot-water fountain that shoots a spray high into the air. Old Faithful geyser in Yellowstone Park in the northern United States sends a jet of water over 60 meters in the air roughly every hour.

Sometimes when magma and pyroclasts explode from vents deep in the interior, the vents and feeder channels are then filled with coarse fragments called *breccia*. The structure that is formed from all this rock debris is known as a *diatreme*. The minerals and rock found in some diatremes are formed at extreme depths in the upper mantle and then forced under pressure to the surface to be ejected explosively.

The surrounding softer soil often erodes from around a hard diatreme and leaves the rock standing like a lone guard. This is the case of a towering structure called, Shiprock, standing in New Mexico in the southwestern United States.

Hot Spots

Volcanoes formed singly, away from plate boundaries, are usually a result of *hot spots.*

A hot spot on the earth's surface is generally found over a mantle chamber of high pressure and temperature magma. For example, the temperature of basalt lava at Kilauea reaches 1160°C.

Volcanoes are formed from vents that rise to the surface from the magma chamber. Island chains like the Hawaiian Islands were formed over a hot spot in the mantle. As the plate moves through the process of the upwelling of new magma within deep ocean rifts, the original volcano overhead moves on past the hot spot. The hot spot then begins forming a new volcano. Repeated eruptions from the hot spot create a string of closely spaced chain of volcanoes.

Currently, a new island called *Loihi* is being formed at the end of the Hawaiian Island chain over the hot spot. Its peak is 1000 meters below the water's surface and will break the surface in the next 10,000 to 100,000 years. Figure 11-2 shows the formation of hot spot volcanic islands.

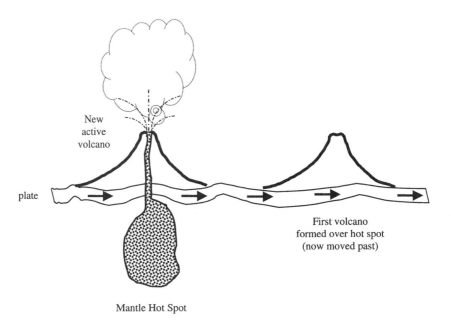

Fig. 11-2.   As the crust moves over a hot spot, a volcano is formed.

Rocks from a string of volcanoes that have passed over a hot spot can be studied and their age calculated. When this is done, the speed and direction of a plate's movement over that spot can be figured out.

Domes

Felsic lavas are so sluggish they barely flow. This type of lava usually makes up a *volcanic dome* globbed out from a central or side vent. Domes are rounded and steep-sided masses of rock that can plug vents and trap gases. When this happens, the pressure builds from below until the dome is blasted to pieces in a violent eruption. This is what happened when Mount St. Helens erupted in 1980.

Scoria Cone Volcanoes

Scoria cone volcanoes, also called *cinder cones*, are the most common type of volcano. Reaching heights commonly less than 300 meters, cinder cones make up the smallest type of volcano.

Scoria cones have straight sides with steep slopes (as much as 35 degrees) and a very wide, often symmetrical peak crater. These volcanoes generally have *Strombolian* type eruptions. They can be seen as single volcanoes in basaltic lava fields or as "piggy-back" cones built up by eruptions along the sides of shield volcanoes and stratovolcanoes. Scoria cones are made up mostly of ejected basaltic *tephra*. This is generally of lapilli size, but can have bomb-size fragments and splatter lava as well. The cinder look of the rock comes from the high amount of gas bubbles (*vesicles*) in the tephra fragments. The shape of a Scoria-type volcano is shown in Figure 11-3.

Fig. 11-3. Scoria or cinder cone volcanoes are the most common type of volcano.

Shield Volcanoes

Shield volcanoes are wide, flat volcanoes with diameters of a few kilometers to over 100 kilometers. They are generally short and squat with heights roughly 1/20th of their width. So a shield volcano that is 100 km across would only be 5 km tall. They have gently sloping lower slopes (2–3 degrees) that build to medium height slopes of about 10 degrees and then flatten at the peak. The spreading shield area grows from the over layering of many hundreds or thousands of fluid lava flows, one upon another, that gurgle and bubble out from the same point of release. Figure 11-4 illustrates the unimpressive low shape and vertical height of shield volcanoes compared to other types.

Shield volcanoes are the largest volcanoes on the planet. Hawaii's volcanoes (like Kilauea and Mauna Kea) are shield volcanoes. In fact, the island of Hawaii is made up of five overlapping shield volcanoes. The Mauna Loa volcano on the big island of Hawaii is the world's largest shield volcano. Shield volcanoes are also said to have *Hawaiian eruptions*, the large spreading volume of lava (thousands of cubic kilometers) that comes from the summit or fissures.

When the flow is almost perfectly symmetrical and of a low volume (10–15 km^3), the eruption produces an *Icelandic shield*. These small volume flows are commonly centered at the summit compared to high-volume linear fissure eruptions typical of the Hawaiian shields. *Galapagos shields* are symmetrical with steep middle slopes (about 10 degrees) and flat tops with large, deep calderas. These shields are thought to be formed by *ring-fracture eruptions* which encircle the caldera and show the site of the caldera collapse.

Shield volcanoes are almost completely basalt, the lava type that is very liquid when it erupts. For this reason, the lava runs out like water from a spilled glass, instead of piling up in a pyramid shape.

Eruptions of shield volcanoes are usually mild and of low-explosive power. The bubbling matter that forms cinder cones and spatter cones around the vent is 90% lava instead of pyroclastic matter. The shield forms around the opening where the volcano spits out a long flowing supply of magma. The hot flowing lava is little changed since it was formed deep in the earth.

Fig. 11-4.   Shield volcanoes are short and squat compared to other types.

Shield volcanoes are common sights in areas of hot spot volcanism. They are also found along subduction arcs or as lone volcanic sentinels in the middle of the ocean.

Composite Volcanoes

Large, long-lived volcanoes of *andesitic*, *dacitic*, and *rhyolitic* composition release a combination of lava flows and pyroclastic debris. These volcanoes, found mostly on the continents, are known as *composite* or *stratovolcanoes*. They usually emit a liquid lava flow, a pyroclastic blast and build up steep conical mounds of alternating layers of pyroclastic material and lava flows. Lava that has hardened within fissures form strengthening, rod-like supports that fortify the cone.

Often the slope of a composite volcano is about 30°, like the angle near the peak of a pyroclastic cone. The base of a composite volcano flattens to about 8–10° angles. This type of volcano is the "classic" volcano shape of the Oregon and Washington volcanoes (Mts. Hood, Baker, Rainier) in the United States and Mount Fugiyama in Japan. Figure 11-5 shows the steep sides and narrow peak crater of a composite volcano.

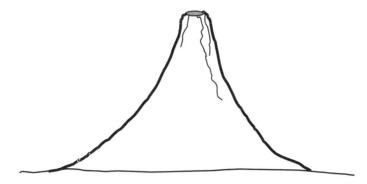

Fig. 11-5.   Composite volcanoes rise majestically above the land around them.

Craters

A bowl-shaped pit or crater is found at the summit of most volcanoes, centered over the vent. During an eruption, the lava blasts or flows out of the vent until the pressure below is released. The lava that is at the tail-end of the eruption cools and sinks back down onto the vent and becomes solid.

When the volcano erupts again, this solid "left over" lava is blasted out in a pyroclastic explosion. The crater gets filled with debris and sometimes the crater sides collapse. When this happens over and over, the crater grows to many times the size of the base vent and hundreds of meters deep. An example of this is Mount Etna in Sicily that is more than 300 meters in diameter and over 850 meters deep.

Calderas

After a massive volcanic eruption lets loose huge volumes of lava from a magma chamber located a few kilometers below the vent, the empty chamber might not be able to support the weight of its roof. When this happens, the volcanic rock above the chamber caves in and a steep-walled basin, larger than the original crater, is formed. This is called a *caldera*. Calderas can be huge, ranging from 4–5 km to more than 50 km across.

Crater Lake, Oregon in the northwestern United States is formed from a caldera that is 8 km in diameter. Geologists think that the original eruption that caused the formation of the caldera happened about 6600 years ago.

A beautiful lake formed within the caldera when water seeped into the crater, created by a composite volcano eruption, was followed by the collapse of the surrounding wall structure. The lake, which is 600 m deep, is the deepest freshwater lake in the United States. Wizard Island, a small pyroclastic cone and lava flow that formed after the collapse of the magma chamber, is located at one end of the lake.

Over time, an old magma chamber sometimes fills with fresh magma, forcing the caldera floor up and into another dome shape. It may stay like this or repeat the eruption and caldera-forming process. When this happens, it is known as a *resurgent caldera*, one that goes through the caldera cycle a number of times.

> A **resurgent caldera** happens when the process of eruption and caldera formation takes place over and over again over time.

Volcanic Gases

Since the tumultuous formation of the Earth, elemental gases have played a part in the composition of the planets' makeup and matter. Volcanologists have collected gases from eruptions and lava flows at great personal risk in

Table 11-3   The poisonous gases spewed from a volcano can be as dangerous as the lava.

Volcanic gas	Percentage of total gases (average)
Water vapor (steam) and carbon dioxide	90–95%
Sulfur dioxide	< 1%
Nitrogen	< 1%
Hydrogen	< 1%
Carbon monoxide	< 1%
Sulfur	< 0.5%
Chlorine	< 0.2%

order to figure out their composition. Table 11-3 lists a few gases associated with volcanic eruptions.

The majority of gas that escapes most eruptions is made up of water vapor. Some of this water comes from groundwater and seawater and some from the atmosphere. Other gases are the result of chemical changes in the magma and rock during melting and upon release to the surface. Gases that may have been trapped deep within the earth may be coming up for the first time since the planet was formed. Can you imagine?

Since most of these gases are poisonous in large quantities, many times people are not killed by pyroclastic fallout and lava when a volcano erupts, but by the searing hot and poisonous gases that make breathing impossible. Archeologists have found that many of the people that died when Mount Vesuvius erupted were killed by the foul volcanic gases released during the eruption.

Climate

When water seeps down to the magma, it turns to steam and can cause a *phreatic eruption*. When this happens, the violence of the eruption is extreme though of lower temperature. Compared to magma eruptions, there is a tremendous amount of steam, but hardly any rock fragments or *ejecta*.

In 1883, Krakatoa in Indonesia exploded with a phreatic eruption. It blew dust over 80 km into the air with a force roughly that of 100 million tons of dynamite. The explosions were heard on Rodriguez Island, 4653 km distant across the Indian Ocean, and in Australia, over 3000 km away and produced orange-red sunsets for several years afterward. Ash fell on Singapore 840 km to the north, Keeling Island 1155 km to the southwest, and on ships that were 6076 km west northwest. Ash was reported over a 750,000 square kilometer area.

The shock of the eruption caused huge waves with breakers over 40 m high, obliterating everything in their path and flinging coral blocks weighing nearly 600 tons ashore. Even though Krakatoa was not inhabited, these breakers caused over 38,000 deaths by flooding low-lying nearby coastal areas and 165 coastal villages were destroyed. When the eruption was over only 1/3 of Krakatoa, previously 45 km^2, was above sea level with newly created northern islands of scalding pumice and ash, where the sea had been 36 m in depth.

The dust from Krakatoa's eruption also caused a decrease in temperature by $\frac{1}{2}$ a degree centigrade worldwide. This cooling effect was reported to have lasted for ten years. When El Chichón in Mexico erupted in 1982, sulfur-containing gases were spewed into the atmosphere blocking sunlight and falling back to the earth over time as acid rain. The same thing happened with the low-ash eruption of Mount Pinatubo in the Philippines in 1991. The curiously cool Northern Hemisphere summer of 1992, following Pinatubo's eruption and discharge of sulfuric gases, caused meteorologists to increase their look at the role of volcanic eruptions in world-wide weather patterns.

Eruption Signals

Volcanoes are fairly "iffy." If we look at all the recorded eruptions over the centuries, we would find that some smoldered for months to years, while others gave no notice at all before belching out with deadly menace. The only reliable predictor of volcanic eruptions seems to be the response of the animal kingdom in the hours before an eruption. Tuned into some sixth or intuitive sense, they seem to know something big is about to happen. But short of keeping your pets nearby all the time, how can humans predict a volcano's misbehavior in time to get clear of the danger?

Geologists use a number of instruments to record the moods of volcanoes and predict their activity. Lasers are used to measure the distance across the crater of a volcano and its growth.

Seismographs measure vibrations in the earth from earthquakes. They detect and locate small earthquakes that geologists correlate with magma movement.

Scientists have used seismographs to identify earthquakes as deep as 55 km beneath Kilauea and Mauna Loa of Hawaii. Earthquakes can signal the movement of magma into vents and channels leading to the surface. Hot magma can take months to move upward. However, when a volcano awakens from a long sleep, there is intense and almost constant seismic activity. When something major is about to happen, seismographs go wild in the days and hours just before an eruption. It becomes clear that an eruption is about to happen when a mob of earthquakes, thousands of them, signal that pressurized magma is splitting solid rock on its driving sprint to the surface.

> Volcanoes are classed into 3 types: **active, dormant, or extinct**.

A volcano is considered *active* if it has erupted within recent recorded history. If a volcano has little erosion and looks fairly fresh, it is considered *dormant* with the ability to become active again at any time. If a volcano has not erupted within recorded time and is eroded to a large extent, it is thought to be *extinct* and very unlikely to erupt again.

However, these are just "rule-of-thumb" activity terms that volcanologists are still working on. The frequency of eruption of volcanoes can be very unique. An average volcano has been found to erupt once every 220 years, but 20% of all volcanoes erupt only about once every 1000 years, and only 2% erupt less than once in 10,000 years. So just because a volcano has been sleeping a very long time, there is no guarantee it will never erupt again!

There are nearly 600 active volcanoes on the planet today. Of these, 17% have been human killers. Roughly 200,000 people have been killed by volcanic eruptions in the past 500 years.

Volcanoes do this in the following nasty ways, either separately or in combination:

(a) Explosions
(b) Fiery ash
(c) Toxic gases
(d) Lava flows
(e) Mud flows or lahars, and
(f) Building collapse.

Luckily, of the 100 highest-risk active volcanoes in the world, only about $\frac{1}{2}$ of them erupt every year. Unlike ancient people, we are a lot more prepared.

Today, volcanologists' increased use of land and space-based technology has improved our odds of predicting and surviving eruptions.

Active volcanoes are those that have erupted within recent (recorded) history. Table 11-4 gives a list of volcanic eruptions that occurred in the past 300 years around the world. The biggest or latest eruptions are given.

Table 11-4   Major volcanic eruptions have been recorded for over 300 years.

Volcano	Location	Eruption year
Mount Fuji	Japan	1707
Oshima	Japan	1741
Cotopaxi	Ecuador	1741
Papadian	Indonesia	1772
Lakagigar	Iceland	1783
Tambora	Indonesia	1815
Galunggung	Indonesia	1822
Krakatau	Indonesia	1883
Santa María	Guatemala	1902
Mount Pelée	Martinique	1902
Vesuvius	Italy	1906
Mount Usu	Japan	1910, 1977
Lassen Peak	United States	1914
Mount Heimaey	Iceland	1973
La Soufriére	Guadeloupe	1976
Nyiragongo	Zaire	1977

(*continued*)

Table 11-4   Continued.

Volcano	Location	Eruption year
Karkar	New Guinea	1979
Mount St. Helens	United States	1980
El Chichón	Mexico	1982
Mauna Loa	United States	1975
Kilauea	United States	1960, 1977, 1990
Nevado del Ruiz	Columbia	1845, 1985
Mount Pinatubo	Philippines	1991
Mount Unzen	Japan	1792, 1991
Galeras	Columbia	1993
Guagua Pichincha	Ecuador	1993

Active volcanoes are classified as Types I through V depending on their activity levels and rate of eruptions.

Geothermal Energy

When a volcano finally stops its activity, the igneous rock in the old magma chamber stays hot for a very long time. Some scientists think maybe even another million years. Groundwater that seeps into the hot rock gets heated and rises to the surface through a fault or fissure where it becomes a *thermal spring*. The temperatures of these springs can get as high as the boiling point of water. The United States has over 1000 thermal springs with even more found throughout the rest of the world.

This is the up side to volcanic activity. The thermal energy rising from deep in the mantle can be channeled for useful means. Reykjavik, Iceland, in the path of the Mid-Atlantic rift, gets its hot water from volcanic springs running just below the surface and heated from below by hot magma.

The United States, Iceland, Russia, Mexico, Japan, New Zealand, and Italy are a few of the countries that tap geothermal steam energy from deeply drilled holes in volcanic rocks to produce electricity for their cities.

Additionally, since ancient times humans have enjoyed the relaxation and appreciated the health benefits of bathing in hot springs. Water temperatures in thermal springs are often hot enough to dissolve minerals from the surrounding rocks. These geothermal fed springs often have high levels of leached minerals that have been used since Roman times for their healing properties.

So, even though volcanoes can be extremely destructive and dangerous, they also provide geothermal energy, soil carbon enrichment, and elemental minerals to surface waters. If we just remember to respect their awesome power, volcanoes can be of much more service than harm.

Quiz

1. Volcanoes can kill in all of the following ways, except
    (a) fiery ash
    (b) lava flows
    (c) loud rumbling
    (d) explosions

2. A pyroclastic flow is a
    (a) popular action toy that bends two directions
    (b) denser-than-air mixture of fine ash and hot gases
    (c) thick viscous type of lava
    (d) lighter-than-air mixture of fine ash and hot gases

3. A volcano that is considered dormant
    (a) will never erupt again
    (b) only erupts every 2000 years
    (c) has not a lot of lava flow
    (d) has the ability to become active again at any time

4. Scoria cones
    (a) are the most common type of volcano
    (b) are seldom found
    (c) are very flat and low
    (d) hold at least two scoops of ice cream

5. Pahoehoe is the Hawaiian word for
    (a) soup
    (b) ropy
    (c) hello
    (d) bad breath

6. The slope of a composite volcano is about
    (a) 15°
    (b) 24°
    (c) 30°
    (d) 45°

7. Tephra is made up of
    (a) a synthetic coating used in cookware
    (b) particles smaller than an ash tuff
    (c) many different types of matter sent blasting out of a volcano
    (d) oak and ash wood

8. The largest pyroclastic particles, larger than 64 mm across, are called
    (a) bombs
    (b) lapilli
    (c) whoppers
    (d) cinders

9. A vent located at the top of a volcano and at the end of a vertical
    feeder pipe is called a
    (a) horizontal vent
    (b) central vent
    (c) sedimentary vent
    (d) air conditioning vent

10. Shield volcanoes are normally short and squat with heights roughly
    what fraction of their width?
    (a) 1/3rd
    (b) 1/4th
    (c) 1/10th
    (d) 1/20th

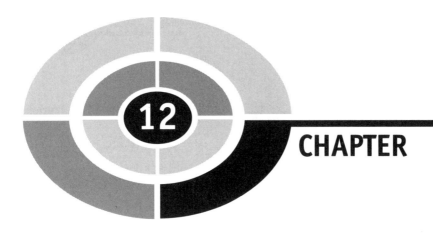

Earthquakes

You're going about your normal daily routine when suddenly, something is not quite right. It may be frighteningly obvious, like watching your shelves emptied by a sudden violent shaking that stops just as mysteriously as it began. It may be very subtle, perhaps just a slight queasy feeling and the gentle swaying of a hanging plant. If you're in an open space and in a good mood, it can even be fun – a free roller coaster ride that lasts a few seconds.

Have you ever been working in an office or at home and suddenly everything goes crazy? The furniture slides a few feet, pictures tilt, glassware clatters and falls, the fish tank sloshes and you wonder if you're just dreaming or in the middle of a Hollywood disaster film. If you live near the margins of an active continental or oceanic plate, neither is the answer. You have just experienced an *earthquake*; a sudden shift in the surface of the earth.

But earthquakes aren't always full of violent special effects, they may be slight and do no more damage than swinging a hanging lamp. In fact, if you're outside, you might feel a few seconds of rolling or nothing at all.

> An **earthquake** is the shaking of the earth that occurs when pieces of the lithosphere suddenly shift.

The term *earthquake* describes the sudden slip of a fault between two plates and includes the ground shaking and radiating seismic energy that is caused by the slip. Volcanic activity and other geologic processes may cause stress changes in the Earth that can also result in an earthquake. Earthquakes take place anywhere in the world, although some areas are more likely to experience an earthquake than others. Earthquakes occur in all types of weather, in all climate zones, in all seasons of the year, and at any time of day, making it impossible to predict with any certainty when an earthquake is likely to occur. The best seismologists (scientists who study earthquakes) can do is to look at the historical record of earthquake activity for a geographical area and use this information to calculate the probability of an earthquake taking place in the near future. Reliable earthquake prediction is still in the future.

An earthquake is simply a sudden ground movement (shift) that people feel on the Earth's surface from a fracture in the crust, known as a *fault*. This shift creates vibrations in the rocks and soil, which are known as *seismic waves*.

> **Seismic waves** are caused by vibrations in the earth caused by cracks or shifts in the underlying rock.

Not all fault movement results in seismic waves strong enough for a person to feel. Minor shifts aren't even picked up by sensitive instruments. However, every shift results in some movement of the Earth's crust along a fault surface.

Infamous Earthquakes

Earthquakes are rarely thought of in a positive light. The only thing most people consider good about an earthquake is that it happened some place else! Earthquakes are not new. They have been around since the landmasses were formed. Ancient humans thought earthquakes were punishment by the gods for bad behavior.

Throughout history, earthquakes caused a huge loss of life and were largely undetected until they hit. The earliest recorded evidence of an earthquake has been traced back to 1831 BC in the Shandong province of China. In 350 BC, the Greek philosopher, Aristotle, noted that soft ground shakes more than hard rock in an earthquake.

The world's most devastating earthquake occurred in 1557 in central China, where most of the local people lived in caves carved from soft rock. These collapsed during the earthquake and killed approximately 830,000 people.

In 1760, British geologist, John Michell, wrote that earthquakes and their resulting energy waves were caused by "shifting masses of rock miles below the surface." Michell also studied and published his theories on magnetization, torsion, and rock density. His most important geological essay, published in 1760, was entitled, *Conjectures Concerning the Cause and Observations upon the Phaenomena of Earthquakes*, which showed a remarkable early knowledge of the strata in various parts of England and abroad.

In 1769, an expedition of Spanish explorer and later Governor of Baja California, Gaspar de Portola, reported an earthquake about 48 km southeast of Los Angeles. De Portola was originally sent to find Monterey Bay, but instead found the Santa Clara Valley and later San Francisco Bay.

On April 8, 1906, the tremendous San Francisco earthquake (8.3 magnitude), which caused over 1000 deaths, struck. It knocked down most of the buildings, but more damage was done by the fires caused by broken gas lines to street lamps than by the earthquake's effects.

The San Francisco earthquake ruptured the upper 430 km of the San Andreas fault from northwest of San Juan Bautista to the junction at Cape Mendocino. The earthquake puzzled geologists of the time with its large, horizontal displacements and great rupture length. In fact, the earthquake's significance and recognition of the faults larger cumulative effect was not fully understood until the introduction of plate tectonics more than 50 years later.

Some sections of the San Andreas fault system crack in large earthquakes, while other parts creep, crack in small to moderate earthquakes, or both. The frictional forces along the fault are proportional to the normal forces across the fault. Active fault slip areas affect directional stresses on smaller creep areas. When some parts crack suddenly and others creep, the direction of the adjusted plate motion is affected.

Every year southern California experiences about 10,000 earthquakes. Most of them are not felt. Or maybe people are so used to feeling minor ground movements that they don't even feel them anymore. Annually, only a few hundred quakes are higher than 3.0 magnitudes and only 20–25 are greater than magnitude 4.0. However, when a big earthquake does hit, the aftershocks create earthquakes of all magnitudes for several months. These shifts and shockwaves represent the settling of the land surface into a new position.

It is a bit like a rock hitting a lake. The initial impact causes ripples outward in all directions. Then, as the ripples hit objects like boulders or the shoreline, other ripples are formed until all the energy from the original impact is used. Then the lake surface goes back to baseline stillness (with a new rock at the bottom).

Worldwide, there are roughly 500,000 detectable earthquakes each year; 100,000 are felt, and of those, roughly 100 cause severe damage.

The largest recorded earthquake in the world was a magnitude 9.5 in Chile on May 22, 1960. The largest recorded earthquake in the United States was a magnitude 9.2 that struck Prince William Sound, Alaska on Good Friday, March 28, 1964.

Earthquakes have also been felt in the central United States. In 1811–1812, strong earthquakes shook the Mississippi Valley's New Madrid fault. Vibrations from these 8+ magnitude earthquakes rang church bells almost 1600 km away in Boston, Massachusetts.

In 1976, another deadly earthquake struck in Tangshan, China, where more than 250,000 people were killed. Table 12-1 lists the largest earthquakes (greater than 8.5 on Richter scale) in the last 100 years.

Plate Tectonics

The deepest earthquakes take place at plate boundaries, where the Earth's crust is subducted into the mantle. This movement can push as deep as 750 km below the surface.

At their edges, plates collide, pull apart, or slip sideways past each other. These plate–plate boundaries and their resulting seismic activity along plate margins have a big effect on regional geology.

Most earthquakes and volcanic eruptions take place along the plate boundaries like the boundary between the Pacific and the North American plates. From Chapter 11, we learned that the plate margin around the huge Pacific Plate, the Ring of Fire, is one of the most active areas for earthquakes and volcanic eruptions.

The East African north–south rift region, a 55 km wide zone of active volcanoes and faulting, stretches for more than 3000 km from Ethiopia in the north to Zambezi in the south. It encompasses an active rift zone, where a continental plate is separating into two plates, pulling away from each other.

Table 12-1   Earthquakes of greater than 8.5 on Richter scale.

Location	Year	Magnitude
Off Ecuador's coast	1906	8.8
Kamchatka	1923	8.5
Banda Sea, Indonesia	1938	8.5
Assam, Tibet	1950	8.6
Kamchatka	1952	9.0
Andreanof Islands, Alaska	1957	9.1
Chile	1960	9.5
Kuril Islands	1963	8.5
Prince William Sound, Alaska	1964	9.2
Rat Islands, Alaska	1965	8.7

Types of Earthquakes

The earth's crust is divided into eight major chunks (plates) and a lot of minor plates. In the Pacific Northwest, there are two main plates that affect earthquake activity significantly. Oregon, Washington, and California sit on the North American Plate. The oceanic, Juan de Fuca Plate, is located in the Pacific Ocean, west of the Pacific Northwest coastline. The margin between these two plates is called the Cascadia Subduction Zone and is found about 80 km offshore. When the Juan de Fuca Plate clashes with the North American Plate, it subducts underneath it down into the earth's mantle. The collision of the two plates produces three different types of earthquakes.

SUBDUCTION ZONE EARTHQUAKES

Sometimes as an oceanic plate is shoved underneath a continental plate it sticks instead of sliding smoothly. This sticking causes stress to build up, which may be released suddenly as a large earthquake.

No large earthquakes have been recorded in the Cascadia Subduction Zone since records began in 1790, but the Cascadia has had magnitude 8–9 earthquakes in the past. If an earthquake took place in the subduction zone, it would probably be centered off the coast of Washington or Oregon where the plates converge. The 1989 volcanic eruption of Mt. St. Helens might be a signal of growing activity in the region.

DEEP EARTHQUAKES

Deep earthquakes take place within the Juan De Fuca Plate as it sinks into the mantle. These earthquakes are found roughly 25–100 km in depth. Because of their great depth, aftershocks are not usually felt. Few earthquakes hit east of the Cascade Mountains. The last one took place in 1965 between Seattle and Tacoma, Washington with a magnitude of 6.5.

SHALLOW EARTHQUAKES

Shallow earthquakes are found within the North American Plate. These earthquakes are thought to be caused by stress forwarded from the Cascadia Subduction Zone into the North American plate. Shallow earthquakes are recorded to depths of 30 km or less and have taken place throughout the northwestern United States. In 1993, an earthquake of magnitude 5.6 occurred inland in the Willamette Valley of Oregon and two earthquakes, magnitudes 5.9 and 6.0, occurred near Klamath Falls, Oregon.

Faults

The sudden slip that is an earthquake takes place after a gradual buildup of stress inside the Earth. Rocks, under enough pressure, reach a breaking point. Think of the weakest link idea, where in weakened places, increasing rock stress causes a break. As the stress in a specific area overwhelms forces holding the rocks together, the proverbial "straw that broke the camel's back" takes place. Something cracks and an earthquake is set off.

> A **fault** takes place along a fracture in a plane of the Earth's crust where slip between the two sides has taken place.

A *fault plane* is used to represent an actual fault or a section of a fault. Faults are generally not perfectly flat, smooth planes, but give geologists a rough idea of the direction and orientation of a fault. The intersection of a fault plane with the Earth's surface, along which a crack takes place, is called the *fault line* or *surface trace*. Faults traces are not always obvious on the surface. Figure 12-1 shows how faults change from a stable original position to a new post-fault position.

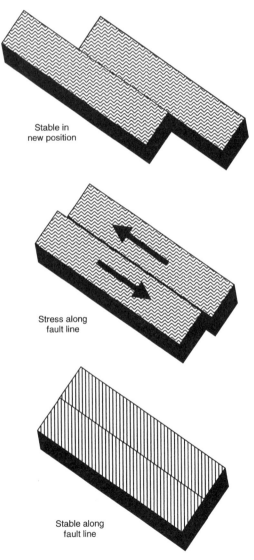

Stable in
new position

Stress along
fault line

Stable along
fault line

Fig. 12-1. Stress builds along a fault until it suddenly snaps into a new location.

A fault trace's *trend* is the direction it takes across the Earth's surface. Trends are used to average out the small bends in a long fault and study its overall direction. This direction is something like a fault's *strike*, but the two are different.

A fault *strike* is the line formed by the intersection of the fault plane with a horizontal plane. The direction of the strike is as angle off north. So when geologists talk about the direction of a certain strike they might call it a northwest-striking fault.

The forces that grind large masses of rock together cause plate margin stress and are unimaginable. That is the reason why the release of pent up seismic energy is equally powerful and damaging.

ST. ANDREAS FAULT

California's active and well-known San Andreas Fault was named in 1895, by geologist A.C. Lawson. The San Andreas Lake, about 32 km south of San Francisco, is said to have given Lawson the idea for the name. However, the San Andreas Fault and its neighboring faults extend almost the entire length of California.

The San Andreas Fault is not truly a single, continuous fault, but a *fault system* made up of many parts. Plate movement is so common in California that earthquakes take place across the zone anywhere, anytime. The San Andreas fault system, over 1300 km long, is also up to 16 km deep so it is no wonder that something is happening all the time.

The average rate of motion across the San Andreas Fault Zone, during the past 3 million years is 56 mm per year. This is estimated to be about the same rate that fingernails grow. At this rate, geologists speculate that Los Angeles and San Francisco will be next to each other in about 15 million years. You will be able to live in San Francisco and go to school or work in Los Angeles in a few minutes (depending on traffic)!

Fault Size and Orientation

Faults can be found in any orientation and at any angle to horizontal (dip), but an active fault's orientation and dip (angle) are not particularly random. These characteristics are affected by the regional stress created by tectonic movement.

Slips take place along some fault planes and angles easier than others. For example, if you are sitting on a flat table, you won't usually slip off unless someone pushes you. However, if someone tips the table, you will slide off, because gravity is stronger than the frictional force keeping you in place. With earthquakes along fault lines, the driving force eventually overcomes the resistant force, causing slip.

Some faults don't break through the surface anywhere along their length. They may have become covered by surface deposition of sediment, but some faults just never reach the surface. Faults may also spread into one or more folds under the surface. A fault that doesn't reach the surface and has no surface trace is called a *blind fault*.

When a blind fault takes place at the surface and forms a chain of hills or a rounded scarp, it is known as a *fold belt* or *fold bend*.

A *scarp* is any roughly linear slope or rock face. A fault that causes a vertical offset will often create a scarp. While some scarps are formed by fault movement, they can also be created by erosion, or other means. When created by faults, scarps are good indicators of surface traces.

ACTIVE FAULTS

Active faults have a high potential for causing earthquakes. *Inactive faults* slipped at some point in time (causing earthquakes), but are now stuck solid. However, if the tectonic processes in an area changes, it is possible for inactive faults to become reactivated.

Faults can measure from less than a meter to over a thousand kilometers in length, with corresponding widths. Large fault depths are limited by the thickness of that portion of the Earth's crust and lithosphere. In southern California, this depth is roughly 15–25 km. Most seismologists study faults that are at least a square kilometer in area, and around a 100 square kilometers in area. Faults of this size or of a greater size crack violently enough to cause significant earthquakes. Geologists estimate there are roughly 200 faults in southern California, considered major faults, and able to produce damaging earthquakes. Smaller southern California faults (thousands to millions) cause only small tremors. Large faults that only crack in a small section along their length cause minor earthquakes.

↑ fault slip area ⇨ ↑earthquake produced

It is important to remember that the size of a fault rupture is directly proportional to the size of the earthquake produced by the slip. Just remember, the bigger the fault slip area, the bigger the earthquake produced. Sometimes, however, when you look at the total surface area of a cracked fault, only a small part of the total area sees a slip. When a set of fractures is large and developed, it is known as a *fault zone*.

Often, you will hear geologists talk about hanging walls and footwalls. These terms refer to the orientation of a horizontal fault in relation to the Earth's crustal position. When the crust is above the plane of the fault, it is said to be a *hanging wall*. It is said that miners could hang their lamps on the wall above in these fault areas. The *footwall* is the part of the Earth's crust below a fault. Miners are said to have walked on the lower footwall side of a mined-out fault.

FAULT DIP

A fault's *dip* or *dip angle* is given by two measurements: angle and direction. The direction is perpendicular to the direction of the strike of the fault plane. The dip angle is the angle of intersection between the fault plane and the horizontal plane.

When recording a dip, geologists might find that a dip slants sharply toward the southwest. If a specific dip angle is given, it is written as an angle off north or south, like "35° east of south," instead of "55° south of east." Because the dip of a plane is always perpendicular to its strike, the dip's direction doesn't have to be given if the strike's direction is noted. Commonly, either dip or strike direction is given.

Finding the strike and dip of a fault plane at the surface can be tough when faults are bound by solid rock. It is more difficult to break a new fault surface through a solid rock than to crack through a previously broken fault. Surface layers of soil and loose sediments, however, are easy to break through. A fault can split through a new and different surface each time.

Another way to find fault strike and dip beneath the surface involves finding the hypocenters of naturally occurring earthquakes and cross plotting these locations. Figure 12-2 illustrates the different parts of a fault compared to the surface.

SLIP RATE

Plate movement forces together or pulls apart the Earth's crust. It causes stress-related dip–slip faults. These faults are found with horizontal and vertical offsets from perpendicular. Their movement permits the crust to thicken or thin in places, while expanding or compressing a rock.

A fault's *slip rate* is the speed that one side of the fault slides in relation to the other. Slip rates are commonly measured in millimeters per year (mm/yr).

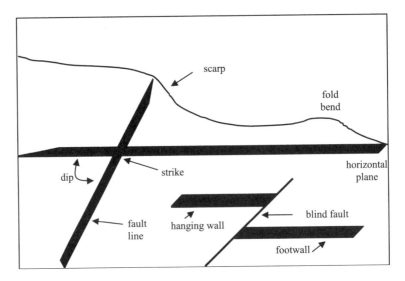

Fig. 12-2. A fault may or may not be seen on the ground's surface.

For example, on the west coast of the United States slip rates range from 0 to roughly 40 mm/yr, though anything over 10 mm/yr is considered speedy. Slip rates of 1–2 mm/yr are thought to be normal for a major, active fault.

It is important to remember that slip rates are averages of total slip along a certain fault over time. They don't slowly move past each other all year long, but instead slip suddenly to a new offset position (a few centimeters apart) in one swift movement.

Slip rates are calculated by using the formula for finding average speed.

$$\text{Average speed} = \frac{\text{Total distance covered}}{\text{Total time required}}$$

Figuring out fault slip rates allows geologists to understand potential danger from faults located near populated areas. Slip rates also provide historical and regional information about faults and their activities over time.

Hypocenters vs. Epicenters

The *hypocenter* of an earthquake is the point at which the earthquake slip starts. Since earthquakes are usually caused by tectonic activity, hypocenters are always located at some depth underground.

> The **hypocenter** of an earthquake is the location beneath the Earth's surface where a fault rupture begins.

If you have even seen television coverage of an earthquake, you probably saw the *epicenter* pointed out on a map. An earthquake epicenter is the point on the Earth's surface directly above the hypocenter. Since this is an easy way to map an earthquake's location, large earthquakes are often named after local geography, such as cities, rivers, or towns closest to the epicenter like the 1906 San Francisco earthquake. The epicenter is often found by comparing the distance from several different measuring stations in a network. The main function of seismograph networks is to find earthquakes. Figure 12-3 shows how the focus of an earthquake radiates and compares to the epicenter.

Although it is possible to find the general location for an earthquake from the information of a single station, it is more precise to use three or more stations. Pinpointing the source of an earthquake helps in evaluating its damage and in relating an earthquake to a geologic site.

> The **epicenter** of an earthquake is the location directly above the hypocenter on the surface of the Earth.

The hypocenter and epicenter may be far apart. A deep hypocenter oriented at an angle away from vertical would not be found along the fault line, but at some distance away. Figure 12-4 illustrates how the epicenter and hypocenter compare to each other for angled faults and blind faults.

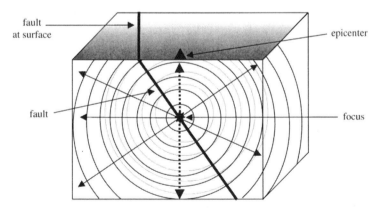

Fig. 12-3. The focus of an earthquake radiates out in all directions.

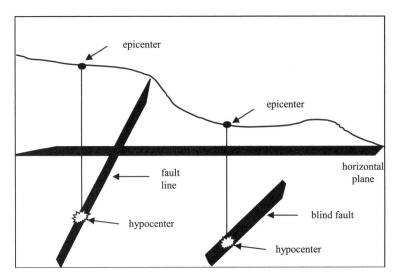

Fig. 12-4.   The epicenter is the surface point above the earthquake's hypocenter.

SEICHE

Have you ever tried carrying a full bucket of water across a distance of 20–30 m? It's tough. Unless you walk very slowly and carefully with little sideways or vertical movement, water sloshes all over the place. The same thing happens after an earthquake. Lakes, swimming pools, fish tanks, and other bodies of water slosh out of their banks in response to the wave affect.

The sloshing of water from a swimming pool or any body of water, after an earthquake is known as a *seiche*. It can go on for a few minutes or up to a few hours; long after the vibrating force has stopped.

To give you an idea of the tremendous force of earthquakes, in 1985, it was reported that the swimming pool at the University of Arizona in Tucson lost water from sloshing (seiche) as a result of the 8.1 magnitude, Michoacan, Mexico earthquake 2000 km away.

Seismic Measurement

When a fault cracks, it causes deformation of surface features. For example, when someone builds a road, railroad, pole line, or fence line across a fault, they may be surprised when they come back after an earthquake, to find everything has changed. The previously straight line is shifted into a shape

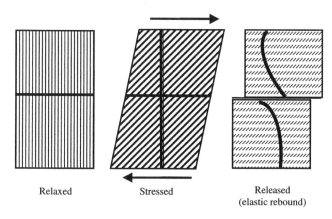

Fig. 12-5. Surface displacement is frequently seen as a result of elastic rebound.

with higher displacement near the fault, a process known as *elastic rebound*. Figure 12-5 shows how this stress is exerted and then rebounds.

EARTHQUAKE INTENSITY

The first known intensity scales used in Europe were used to compare the vibrational effects of one earthquake to another. An intensity scale helped to estimate the overall severity of any one earthquake, compared to others that had taken place previously.

By the early 20th century, the two most commonly used standards of intensity were the *Rossi-Forel* and the *Mercalli* scales, which used Roman numerals. Intensities were rated from I to X on the Rossi-Forel scale, and from I to XII on the Mercalli scale. Higher numbers were equal to stronger earthquakes.

In 1931, the *Modified Mercalli Intensity Scale* was described by Harry Wood and Frank Neumann. It measured the destructiveness of an earthquake and used the same I to XII scale as the original Mercalli scale. The 1906 San Francisco quake was given a XI on the Modified Mercalli Intensity Scale.

SEISMOGRAPHS AND SEISMOGRAMS

Seismology is based on the measurement and observation of ground motion. Originally, people judged an earthquake's strength by the amount of damage it caused.

The first pendulum *seismograph* used to measure the shaking of the ground during an earthquake was developed in 1751, but it wasn't until 1855 that geologists realized faults were the source of earthquakes. The first seismographs using time and motion indicators were built in the late 1800s. Long before electronics, scientists built huge *spring-pendulum seismographs* to measure long-period quake motion. A seismograph, three stories high, in Mexico City is reportedly still in use.

> A **seismograph** is an instrument that records seismic waves (vibrations) onto a tracing called a **seismogram**.

RICHTER MAGNITUDE SCALE

Seismographs record a zigzag trace that shows the changing amplitude of ground movement beneath the instrument. Sensitive seismographs greatly magnify these vibrations and can detect earthquakes all over the world. The time, location, and magnitude of an earthquake can be found from the information gathered by seismograph stations. Figure 12-6 shows a typical trace from a seismograph.

The *Richter Magnitude Scale* was developed in 1935, by Charles F. Richter of the California Institute of Technology, as a math tool to compare earthquake sizes. He got the idea from his early studies of astronomy. He knew astronomers gave stars a magnitude level based on their brightness. By using high-frequency data from nearby seismograph stations, Richter came up with a way to measure the magnitude of an earthquake. This became known as the Richter local magnitude (M_L).

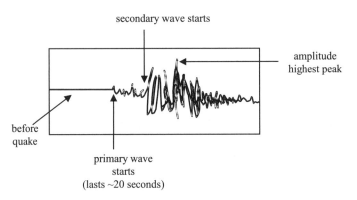

Fig. 12-6. Seismographs record earthquake primary waves, secondary waves, and wave amplitude.

The equation for Richter Magnitude is shown below:

$$M_L = \log_{10} A \text{ (mm)} + \text{(Distance correction factor)}$$

A is the amplitude, in millimeters, measured directly from the photographic paper record of a seismograph. The distance correction factor comes from a table in Richter's *Elementary Seismology* book.

Earthquake magnitude is found from the logarithm of the amplitude of waves recorded by seismographs. Adjustments are included for the distance differences between the several seismographs and the earthquake epicenter. On the Richter scale, magnitude is written in whole numbers and decimals. It is an open-ended scale (10 is not the highest number).

Because the scale is logarithmic, each whole number increment in magnitude actually represents a tenfold increase in amplitude or energy.

Earthquakes with a magnitude of 2.0 or less are often called *microearthquakes*. Most people don't feel them and only nearby seismographs record their movement. Earthquakes of 4.5 or greater magnitudes (thousands each year) are strong enough to be recorded by sensitive seismographs worldwide.

Large earthquakes, such as the 1964 Alaskan earthquake, have magnitudes of 8.0 or higher. On average, one earthquake of 8.0 occurs somewhere in the world every year.

Seismic Waves

Today's digital seismic monitoring instruments use waveforms to record and analyze seismic data. They give information not only on local seismic activity, but also earthquakes at far distant locations. When the Chilean earthquake took place in 1960, seismographs recorded the seismic waves that traveled all around the Earth. Seismic waves rattled the Earth for days. This occurrence is known as the *free oscillation* of the Earth.

BODY WAVES

There are two types of seismic waves, *body waves* and *surface waves*. Body waves travel through the interior of the Earth, while surface waves travel only within the top surface layers. Most earthquakes take place at depths of less than 80 km below the Earth's surface. There are two types of body waves, *P waves* and *S waves*. P waves bunch together and then spread

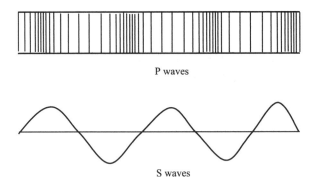

P waves

S waves

Fig. 12-7.   P waves and S waves have different speeds and wave actions.

apart when they move. It is a bit like the movement of an inch worm or a slinky. S waves are like rolling ocean waves or like when you snap a rope. The oscillations are in a waveform. Figure 12-7 gives you a rough idea of how P waves and S waves look.

Primary waves

P waves are the fastest seismic waves and the first to arrive at any monitoring station. P waves travel at about 5 km/s (14 times faster than sound waves travel through air). For this reason, they are called *primary waves* or *P waves* of an earthquake.

P waves are *longitudinal compressional waves*. Like sound waves, they travel through rock or buildings causing squeezing and expanding, parallel to the direction of transmission. The speed that P waves travel depends on the types of matter through which they move. Generally, in the case of P waves, the denser the matter, the faster the wave travels.

Because P waves are the first seismic waves to reach any given location after an earthquake takes place, their arrival can easily be identified on a seismogram. The tracing is flat during zero or background vibration, then jumps into action as the first P waves arrive.

Secondary waves

S waves arrive at a monitoring location second, since they only travel around 60% of the speed of P waves. For this reason, they are called *secondary* or *S waves*.

S waves are *transverse shear waves*. They cause a shearing, side-to-side motion transverse (perpendicular) to their direction of travel. Because of this, they can only travel through a substance like rock that has shear strength. Liquids and gases have no shear strength and S waves cannot travel through water, air, or even molten rock.

Since the seismogram is already jumping wildly with the action of the P waves, it can be tricky to figure out when S waves arrive. However, seismologists study the amplitude and wavelength of a seismic recording. S waves are often lower in frequency and longer in wavelength. So a sudden increase in wavelength tips off the arrival of an S wave. A better indicator of an S wave's arrival is a sudden jump in the amplitude. In cases where the earthquake is large and the source is nearby, however, this doesn't always work because the P wave movement has not yet slowed to the point where the S wave arrival overwhelms it.

SURFACE WAVES

Surface waves are also divided into two types, *Love waves* and *Rayleigh waves*. Each of the two surface waves travels literally along the surface of the Earth and produces a distinct type of motion.

Love waves produce motion that is perpendicular to the direction of wave travel in a horizontal orientation *only*. This is the type of horizontal shearing that wipes out building foundations.

Rayleigh waves produce a rolling motion like the waves on the sea. An object like a building on the surface will experience both an up-and-down, bobbing motion transverse to and a back-and-forth motion parallel to, the Rayleigh wave direction of travel. The two movements combine to create a rolling, elliptical action that is extremely difficult for buildings not specially designed for earthquakes to withstand.

Love and Rayleigh surface waves are responsible for a lot of damage that takes place during earthquakes. They are not as basic as P and S waves, but create a lot of problems anyway.

SEISMIC MOMENT

Seismologists have developed a standard magnitude scale that is independent of the instrument type. It is called the *moment magnitude scale* and comes from calculations of *seismic moment*.

Seismic moment is figured out through the physics of *torque* (twisting). A torque is a force that changes the angular momentum of a system. It is defined as the force times the distance from the center of rotation. Earthquakes are caused by internal torques, which are caused by stress interactions of tectonic plates and faults. Seismologists calculate moment magnitude with the following factors:

$$\text{Moment} = \text{rock rigidity} \times \text{fault area} \times \text{slip distance}$$

MAGNITUDE AND INTENSITY

Magnitude and *intensity* measure different earthquake characteristics. Magnitude measures the energy released at the source of the earthquake and is determined by seismographs measurements. The magnitude is the same no matter where you are, or how strong or weak the shaking at various locations.

> The **magnitude** of an earthquake is a measured value of an earthquake's size.

In Richter magnitude, the amount of movement (amplitude) caused by seismic waves determines its magnitude.

Intensity measures the strength of shaking produced by the earthquake at a certain location. Intensity is determined from effects on people, human structures, and the natural environment.

> The **intensity** of an earthquake is a measure of the shaking at a certain location created by the earthquake.

Table 12-2 gives intensities that are often seen near the epicenter of different earthquakes' magnitudes.

Using Richter's open ended scale allows more accurate reports of earthquake intensity. It doesn't depend on people's feelings and observations like earlier scales did.

Today with the Internet and more and more households submitting reports, it is possible for intensity maps to be created within hours of a large earthquake. These intensity maps help seismologists figure out how subsurface matter relates to what is felt in an earthquake and the damage to buildings and homes.

Table 12-2   The amount of damage from earthquakes varies with intensity.

Magnitude	Intensity	Effects
1.0–3.0	I	Rarely felt
3.0–3.9	II–III	II – only felt by those lying still and in tall buildings III – felt by people indoors and in tall buildings, but not recognized as an earthquake; cars may rock a bit; vibrations like truck going by; time estimated.
4.0–4.9	IV–V	IV – felt indoors by many, outdoors by few during the day; some awakened at night; dishes, windows, and doors jarred; walls crack and pop; sounds like heavy truck ramming building; cars rocked a lot V – felt by almost everyone; many wake up; dishes and windows broken; objects tipped over
5.0–5.9	VI–VII	VI – felt by everyone, many alarmed; heavy furniture moved; some fallen plaster; overall damage slight VII – damage minor in buildings of good design and construction; slight to medium in well-built structures; lots of damage in poorly built or designed structures; some chimneys broken
6.0–6.9	VIII–IX	VIII – damage minor in specially designed structures; lots of damage and some collapse in common buildings; huge damage in poorly built structures, chimneys, factory stacks, columns, monuments, and walls fall; heavy furniture flipped IX – damage great in specially designed structures; well-designed frame structures jerked sideways; great damage in substantial buildings, with partial collapse; buildings moved off foundations
7.0 and higher	X and higher	X – some well-built wooden buildings destroyed; most masonry, frame, and foundation destroyed; rails bent XI – few masonry structures left standing; bridges destroyed; rails bent completely XII – damage complete; lines of sight and level are distorted; objects airborne

Table 12-3   Earthquake totals of the past 100 years.

Time frame	Number of earthquakes
1900–1910	232
1911–1920	198
1921–1930	166
1931–1940	212
1941–1950	314
1951–1960	192
1961–1970	211
1971–1980	193
1981–1990	106
1991–2000	173
2001–2003	44
Total	2041

EARTHQUAKE PREDICTIONS

Since millions of earthquakes (most very minor) hit somewhere on the planet every year, there is no way they can all be predicted or even recorded. Most people want to know when to expect the "Big One."

In fact, some people even think there is an earthquake season. There is no such thing. Statistically, earthquakes take place all the time in all kinds of weather. Season has no affect on rock stresses deep in the earth. Barometric pressure changes are nothing compared to plate tectonic pressures. Table 12-3 lists the number of earthquakes and their numbers over the past hundred years in the United States.

But just like real estate, location is everything. Alaska has the most earthquakes and is one of the most seismically active area in the world. Alaska has a magnitude 7 earthquake almost every year and a magnitude 8 or greater earthquake about every 14 years.

In the continental United States, only Florida, Iowa, North Dakota, and Wisconsin had no recorded earthquakes between the years of 1975–1995. So, if you really don't like earthquakes, these states might be on your "move to" list.

Though geologists are beginning to get a handle on tectonics, faulting, and earthquake processes, earthquake predictions are still fairly uncertain.

On February 4, 1975, the Chinese government issued an immediate earthquake warning to the area of the city of Haicheng and began a huge evacuation effort. Haicheng experienced an earthquake of magnitude 7.3, about nine hours later. Fortunately most of the population was outside when 90% of the city's buildings were severely damaged or destroyed. Thankfully, injuries were few.

Although there have been other prediction successes in this region of China, the misses are frequent. On July 28, 1976, a 7.8 magnitude earthquake hit the Chinese city of Tangshan, 150 km east of Beijing, and home to over one million people. The hypocenter was located directly under the city, at a depth of 11 kilometers. About 93% of all buildings in the city were destroyed, and approximately 240,000 people were killed. The earthquake came as a complete surprise to Chinese seismologists, even though the area was monitored constantly.

EARTHQUAKE SWARMS

Earthquakes sometimes occur in small clusters or *swarms* with no really big jolt. The larger vibrations in a swarm are all about the same magnitude. Swarms take place in a small area and can last a day or months.

In some ways, swarms resemble aftershocks. Aftershocks from a main earthquake are thought to come from the sudden stress loads placed upon the rocks near a fault rupture.

Swarm patterns are different. Geologists believe geothermal energy in the crust and areas of high heat flow play a role in swarm mechanisms. After a main shock, heat allows rocks to release stress more easily. Areas of high heat flow are subject to long swarms that die down slowly.

Other ideas about earthquake swarms include:

- the formation of new faults
- fault creep
- increase in the pressure of fluids at depth, and
- sub-surface movements of magma.

Earthquake prediction could become a reality. Just as the National Weather Service predicts hurricanes, tornadoes, and other bad storms, the National Earthquake Information Center may one day release earthquake forecasts. The United States Geological Survey and other federal and state agencies, as well as universities and private institutions are also working on the problem.

TSUNAMI

A *tsunami* is a series of waves caused by large earthquakes or landslides at or beneath the sea floor. This seismic movement with the displacement of huge volumes of the sea water above it creates large, fast moving waves. A tsunami results in a coastal earthquake or from an earthquake in a far distant part of the ocean. Coastal areas may suffer little damage from an oceanic earthquake, but can be ruined by the resulting tsunami.

Although both are sea waves, a *tsunami* and a *tidal wave* are two different unrelated processes. A tidal wave is produced by high winds, while a tsunami is caused by an underwater earthquake or landslide (usually triggered by an earthquake) displacing ocean water.

Geologists have learned a lot about the wild tectonic mechanisms that happen within the Earth. As better instrumentation and understanding develops, earthquakes may lose their element of surprise!

Quiz

1. Seismic waves are caused by
   (a) ballroom dancing
   (b) loud cars
   (c) water pressure
   (d) vibrations in the earth from shifts in the underlying rock

2. The fastest seismic waves and the first to arrive at any monitoring station are
   (a) P waves
   (b) Q waves
   (c) R waves
   (d) S waves

3. Earthquakes that take place in bunches with no main quake are called
   (a) faults
   (b) swarms

(c) stacks

(d) inverted

4. The Richter Magnitude scale
   (a) measures tidal waves
   (b) rates earthquakes on a 1–12 scale
   (c) is an open ended scale
   (d) is never used

5. Because of their side-to-side motion, S waves are sometimes called
   (a) bobbing waves
   (b) transverse shear waves
   (c) primary waves
   (d) simple waves

6. The sloshing of water, caused by earthquake vibrations, from a swimming pool or body of water is known as a
   (a) seiche
   (b) spill
   (c) washout
   (d) sieve

7. Earthquake swarms are thought to be caused by all of the following, except
   (a) the formation of new faults
   (b) fault creep
   (c) increased pressure of fluids at depth
   (d) extremely cold winters

8. A series of waves caused by large earthquakes or landslides at or beneath the sea floor is called a
   (a) beach bash
   (b) roller
   (c) tsunami
   (d) tidal wave

9. What word describes the sudden slip on a fault as well as surface shaking and radiating seismic energy?
   (a) plate tectonics
   (b) earthquake
   (c) tsunami
   (d) fault bend

10. The location directly above the hypocenter on the surface of the earth is called the
    (a) scarf
    (b) blind fault
    (c) epicenter
    (d) geologic zone

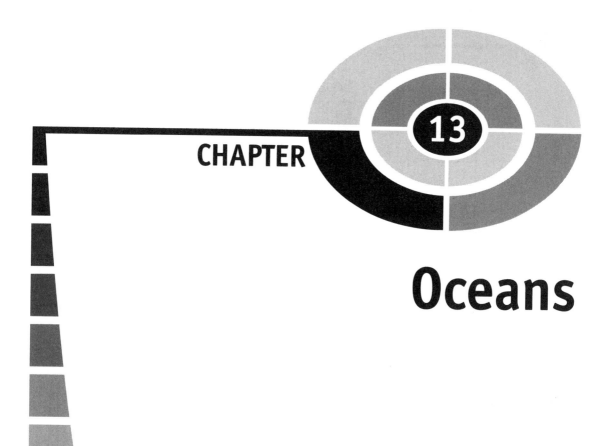

CHAPTER 13

Oceans

Since ancient times, the oceans have been a place of mystery and interest. These deep blue expanses have given us food, trade items, trade routes, adventure, and entertainment. They have separated us and brought us together. From surfers to sand castle builders, people are drawn to ocean shores. Unlike other areas of the Earth, the ocean's treasures are more accessible. Everyone wants to see the latest treasure that has washed up from far off places.

The motion picture industry has used the oceans as subject matter and backdrop for years. Everything from *Titanic* to *Finding Nemo* has captured audiences' imaginations with their possibilities. Table 13-1 gives an idea of the many creative ways the ocean has been portrayed in films.

Landlocked people of the world have gotten their best glimpse into the adventure and mystery of the world's oceans, through the work of French oceanographer and undersea biologist, Jacques Cousteau and his sons. Cousteau's curiosity of the world's oceans opened up an unseen world to everyone that was probably experienced only by a select few.

Cousteau invented the equipment that made ocean exploration possible. *SCUBA* (self-contained underwater breathing apparatus) and the iron lung

Table 13-1   The world's oceans have been a part of many memorable films.

Movie title	Year
Mutiny on the Bounty	1935
20,000 Leagues Under the Sea	1954
Beach Blanket Bingo	1965
Jaws, Jaws 2, Jaws 3-D, Jaws: the Revenge	1975, 1978, 1983, 1987
The Poseidon Adventure	1979
Splash	1984
The Abyss	1989
The Little Mermaid	1989
Hunt for Red October	1990
Free Willy, Free Willy 2	1993,1995
Waterworld	1995
Titanic	1997
Deep End of the Ocean	1999
The Perfect Storm	2000
Cast Away	2000
U-571	2000
Blue Crush	2002
Pirates of the Caribbean	2003
Finding Nemo	2003
Master and Commander: The Far Side of the World	2003

allowed divers to reach depths previously impossible. Today, SCUBA diving is probably at an all time high as a recreational sport. More and more people are fascinated by the complex and beautiful world beneath the waves.

Founder of the French Navy's Undersea Research Group in 1946, Cousteau became commander of the research ship *Calypso* (a converted minesweeper) in 1950 and most of his undersea films were made with the *Calypso* as his base of operations; a total of over 30 voyages. His films include *The Silent World* (1956) and *World Without Sun* (1966). Both won Academy Awards for best documentary. Cousteau's books include *The Living Sea* (1963), *Dolphins* (1975), and *Jacques Cousteau: The Ocean World* (1985).

In 1974, Cousteau started the Cousteau Society to educate, protect, and preserve the world's oceans. The membership of this nonprofit group includes more than 300,000 members worldwide. Cousteau was awarded the Medal of Freedom by United States President Ronald Reagan in 1985. In 1989, Cousteau was honored by France with membership in the French Academy.

A lifelong voice of undersea conservation, Jacques Cousteau died at 87 on June 25, 1997. He left behind a legacy of wonder and hope for the world's oceans that he shared with children and adults worldwide.

Yet with all the sophisticated instrumentation that scientists and oceanographers possess today, the oceans still hold many secrets.

Nearly 72% of the Earth's surface is covered by oceans. The oceans hold $1,300,000,000 \, km^3$ of water. So it should come as no surprise that the interaction between climate and the world's oceans has fascinated scientists for years.

It is thought that the first life developed in the oceans as microorganisms. The oceans have served as cradle, restaurant, and recreation throughout the Earth's history. Oceans seem to be a realm all their own. Even now, when continents have been mapped and their interiors made accessible by road, river, and air, the oceans are largely a mystery. Figure 13-1 shows the different depths to which the oceans have been explored.

Salinity

It has been estimated that if all of the oceans' water were poured off, salt would cover the continents to a depth of 1.5 m. That is a lot of salt!

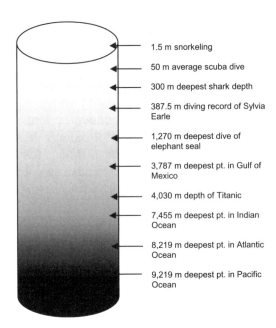

Fig. 13-1. The oceans of the world have different depths.

Salinity is the amount of salt found in 1 kg of water. Salinity, or salt content, is written in parts per thousand (ppt) because there are 1000 g in 1 kg.

The average ocean salinity is 35 ppt. This number varies as rainfall, evaporation, river runoff, and ice formation changes it slightly (32–37 ppt). For example, it is said that the Black Sea is so diluted by river runoff, its average salinity is commonly around 16 ppt.

Freshwater salinity is usually less than 0.5 ppt. Water between 0.5 ppt and 17 ppt is called *brackish*. In areas where fresh river water joins salty ocean water, like estuaries, the water becomes brackish.

When salt water gets to the polar regions, it cools and/or freezes, getting saltier and denser. Cold, salty water sinks. The level of ocean salinity increases by depth. It is divided into three vertical layers. The surface layer has a mixed salinity depending on rainfall or runoff from the land. The middle layer is called the *halocline* with a medium range of salinity. The deepest and coldest ocean water has the highest level of salinity.

Fish and animals that live in seawater have worked out ways to survive in a salty environment. Most marine creatures are able to maintain nearly the same concentration of salinity within their bodies as the surrounding environment. When they are moved to an area of much less salinity, they die. You can't put a saltwater fish in a freshwater aquarium!

Temperature

The ocean has a broad temperature range from warm (38°C) shallow coastal waters of the equator to the nearly freezing arctic waters.

The freezing point of seawater is about −2°C, instead of the 0°C freezing point of ordinary water. Salt lowers the freezing point of seawater. As seawater increases 5 ppt in salinity, the freezing point decreases by −17.5°C.

The ocean is also divided into three vertical temperature zones. The top layer is the *surface layer* or *mixed layer*. This warmest layer is affected by wind, rain, and solar heat. Have you ever been swimming in a deep lake? As you get farther away from the sun-heated surface, it gets colder. Your feet are colder than your upper body.

The second temperature layer is known as the *thermocline* layer. Here the water temperature drops as the depth increases, since the sun's penetration drops too.

The third layer is the *deep-water layer*. Water temperature in this zone sinks slowly as depth increases. The deepest parts of the ocean are around 2°C in temperature, with inhabitants that either like very cold water, or have found specialized environments like a volcanic vent that heats the water dramatically. Figure 13-2 illustrates the thermocline's location in the ocean's temperature layers.

Density

Temperature, salinity, and pressure come together to influence water *density*, which is the weight of water divided by its volume. Cold seawater is denser than warm, coastal water and will sink below the less dense layer.

$$\text{temperature} + \text{salinity} = \text{density}$$

The ocean waters are similarly divided into three density zones. Less dense waters form a surface layer. The temperature and salinity of this layer varies according to its contact with the air. For example, when water evaporates, the salinity goes up. If a cold north wind blows in, the temperature dips and that also affects density.

The middle layer is the *pycnocline* or *transition zone*. The density here does not change very much. This transition zone is a barrier between the surface zone and the bottom layer, allowing little to no water movement between the two zones.

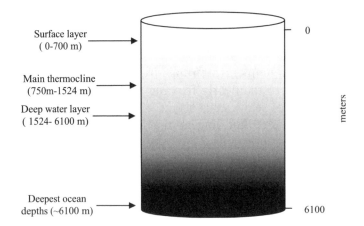

Fig. 13-2. The oceans have distinct temperature zones.

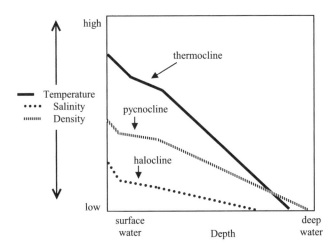

Fig. 13-3. The continental shelf is an extended shallow area of ocean just off the shoreline.

The bottom layer is the *deep zone*, where the water stays cold and dense. The polar regions are the only places where deep waters are ever exposed to the atmosphere because the pycnocline is sometimes not present. Figure 13-3 shows the three-transition layers of density (pycnocline), salinity (halocline), and temperature (thermocline) according to depth.

Where a river meets the ocean, fresh water flows into salty water. The colder river waters are often less dense than the warmer ocean waters. The density differences can create distinct layers in the relatively shallow waters.

A submarine or diver operating in this environment and wanting to stay at the same horizontal level would have to alter buoyancy to adjust for density changes. The same thing happens when a river flows into the ocean. It drags along a portion of sediment into the ocean. This changes the shape of the ocean floor by sedimentation and silting.

Pressure

Although no one really thinks about it, air pushes against us at a constant pressure. At sea level, this pressure is 14.7 pounds per square inch (psi), or $1 \, kg/cm^2$. Our body handles this constant push, by pushing back with the same amount of force. On the top of a mountain, the pressure is less.

But water is a different story. Water is a lot heavier than air. The pushing force (pressure) goes up when you enter the water. In fact, at 10 m in depth, one atmosphere (14.7 psi) pushes down on you.

It's possible for humans to dive three or four atmospheres with the right scuba equipment, but to go any deeper, tough pressurized vehicles like research vessels and submarines are needed.

Ocean citizens like whales seem to be unaffected by pressure shifts. They chug along in the ocean and dive through rapid pressure changes, all the time, without even thinking about it. Although who really knows what a whale thinks about?

Sperm whales are known to dive to depths of 2250 m and can stay down for over an hour. The pressure change from the surface is more than 220 atmospheres! Scientists are still trying to figure out how they do it.

Acoustics

Though Jacques Cousteau titled a book about the sea, *The Silent World*, the oceans are pretty noisy in places. Where currents smash up against rocky cliffs, the turbulence fills the ocean with sound. However, human ears aren't sensitive enough to hear all the different frequencies.

Water is a great sound conductor. It doesn't absorb sound, but allows it to travel great distances before it fades away. The speed of sound through the water is 1450 to 1570 m/s. The travel time increases as the water temperature increases.

Ocean animals are fairly loud. They chatter and call to each other while swimming, squeak when scared, brag when they find food, whistle to send out warnings, ping to inspect their surroundings, and sing to each other on Valentine's Day. Well, they sing anyway.

Dolphins and whales use a method called *echolocation*. By emitting a series of clicks and whistles and then listening for the echoes of the sounds bouncing off objects, they can tell where things are. They do this to accurately locate fish, turtles, logs, boats, reefs, and whatever else is around. From the direction and strength of the echo, dolphins and whales get a mental image of their environment. By echolocation, they can tell the size, distance, and direction of objects in their path.

> **Echolocation** is the method used by whales and dolphins to find out what is happening in the ocean around them.

Sonar is also used by oceanographers to study the ocean floor. It works a lot like echolocation. By sending out signals and picking up the echoes, scientists can get a picture of the features of the ocean floor. This is how shipwrecks like the *Titanic* were located. Fishermen also use sonar to find large schools of fish.

Oceanographers discovered one part of the ocean that has different and better acoustics than other parts of the ocean. It is known as the *SOFAR channel*, which is short for sonic fixing and ranging channel. In the SOFAR channel, low-frequency sounds can travel for hundreds of kilometers very well. If you rise or go deeper vertically the sound fades much faster. Scientists think that this is the main "long distance telephone line" that whales use to communicate to each other over the ocean's expanse.

Optics

Most of the organisms in the ocean depend on sunlight. Plants and bacteria, such as *kelp*, *sea grass*, and *plankton*, use sunlight to make energy by *photosynthesis*. These organisms provide food for fish and some whales. Fish are eaten by larger fish, sharks, and other predators.

This is the food chain of the sea. Sunlight is the energy and heat source for the ocean's food chain. Surface heat and thermal currents that bring in nutrients make it possible for animals to live in warmed ocean waters.

As sunlight penetrates the ocean, it gets absorbed. Like the other ocean gradients we have seen in temperature, salinity, and density, ocean waters can be divided into three vertical regions based on the amount of sunlight that penetrates.

The first zone, or *euphotic zone*, starts at the water's surface and dips downward to about 50 m in depth. This depends on the time of year, the time of day, the water's transparency, and whether or not it is a cloudy day. This is the ocean region where there is still enough light to allow plants to carry on photosynthesis. All plankton, kelp forests, and sea grass beds are found in the euphotic zone. Figure 13-4 shows how these optical regions stack up.

The next zone is the *dysphotic zone*, which reaches from around 50 m, or the edge of the euphotic zone, to about 1000 m in depth. In this zone, there is enough light for an organism to see, but it is too dim for photosynthesis to take place. When divers go deeper into the dysphotic zone, they experience less and less light with depth.

When you get to the *aphotic zone*, there is no light. The aphotic zone extends from about 1000 m of depth or the lower edge of the dysphotic zone to the sea floor. For many years, scientists didn't think there were any animals that lived in this zone, but with deep diving scientific submarines, they have found that several specialized species do exist.

In 1977, geologists who had been exploring ocean fractures discovered booming thermal volcanic vent communities living without sunlight on the barren sea floor. These big, alien-looking creatures used a previously unknown energy process that doesn't include solar heat.

Scientists discovered that the food chain depends on sulfur for energy, not sunlight in vent communities. Deep ocean bacteria transform the chemicals they get from this high-sulfur environment to energy. This energy transformation process is called *chemosynthesis*.

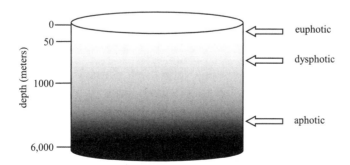

Fig. 13-4. Sunlight depth creates different zones in ocean waters.

Other dark-living animals eat bacteria, shelter bacteria in their bodies, or consume bacteria-eaters in the chain. Vent worms, for example, have no mouth or digestive tract. Instead, they maintain a symbiotic relationship with these chemosynthetic bacteria. The bacteria live in their tissues and provide them with food. Stranger still, scientists found that hemoglobin, which transports hydrogen sulfide to the bacteria, gives vent worms a red color.

Black smokers, the hottest deep ocean hot springs, have been known to reach temperatures of 380°C. This extremely hot water, mixed with hydrogen sulfide and other leached basaltic trace minerals, is emitted from fractures in the earth's crust. *White smokers* have a different composition and lower temperatures.

Bioluminescence

Some animals in the aphotic zone create their own light through a chemical reaction. This is called *bioluminescence*. It is a lot like the reaction that fireflies perform in their warm summer evening dances on land.

These microscopic organisms, floating on the surface, produce their own light through bioluminescence. Disturbances by boats, ships, and swimmers can all cause these organisms to glow. This is an awesome sight at night! Because of these glowing party animals of the sea, the wakes of passing ships have been seen to last for over 10 km!

Ocean Regions

Surrounding nearly all continents is a shallow add-on to the land called the *continental shelf*. This land shelf is fairly thin, 30–60 m deep, compared to the thousands of meters deep in the open ocean. It stretches to the *continental slope* where the deep ocean drops away.

The continental shelf is formed from sedimentary erosion of the land plates and washed into the ocean by rivers and waves. This nutrient-rich sediment provides food for microscopic plants and animals at the beginning of the food chain as we saw earlier.

The amount of nutrients is so plentiful on the continental shelf that great schools of fish, such as tuna, cod, salmon, and mackerel, thrive in busy communities here. The world's continental shelf regions also contain the highest amount of *benthic life* (plants and animals that live on the ocean floor).

The continental slope connects the continental shelf and the ocean's crust. It begins at the *continental shelf break* where the bottom sharply drops off into a steep slope. It commonly begins at 130 m depth and can be up to 20 km wide. Figure 13-5 illustrates how the continental shelf slopes off to the deeper ocean bottom.

The continental slope, counted as part of the main landmass, together with the continental shelf is called the *continental margin*. Undersea canyons cut through many of the continental margins. Some of these are sliced out by *turbidity currents*, which drive sediments across the bottom.

Past the continental slope, we find the *continental rise*. As currents flow along the continental shelf and down the continental slope, they pick up, carry, and then drop sediments just below the continental slope. These sediments buildup to form the wide, gentle slope of the continental rise.

The deep ocean basin, located at a depth of about 3.7–5.6 km deep, covers 30% of Earth's surface and has such features as, *abyssal plains*, deep-sea *trenches*, and *seamounts*.

The abyssal plain is the flat, deep ocean floor. It is almost featureless because a thick layer of sediment covers the hills and valleys of the ocean floor below it. Deep-sea trenches are the deepest parts of the ocean. The deepest one, the Mariana Trench in the South Pacific Ocean, is nearly 11.0 km deep.

SHORELINES

Waves, various currents, and tides all intermingle with specific land, rocks, and plates to give shorelines unique characteristics.

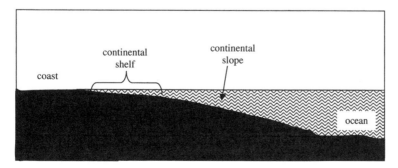

Fig. 13-5.   The continental slope stretches between the continental shelf and the ocean floor.

To most of us, a day at the beach is a day of water, sun, surf, and sand. Beaches are composed mostly of sand, pebbles, and rocks depending on the climate and nearby land mass. However, they are constantly changing. Waves and wind are the endless forces that never stop. They build beaches up and wash them away. The main factors that affect the creation and maintenance of shorelines all around the world include:

- Rising of coastal area with associated erosion
- Sinking of coastal areas with sediment deposition
- Types of rocks or sediments present
- Changes in sea level
- Common and storm wave heights, and
- Heights of tides affecting erosion and sedimentation

Some beaches go on and on for many kilometers, while others are called pocket beaches, with just a nick of sand or smooth stones in a long shoreline. Sometimes beaches are backed by rocky cliffs, while other times they are flanked by sand dunes.

Beaches are generally composed of three parts: *backshore*, *foreshore*, and *offshore*. The backshore extends from the dunes or land margin to the surf. The foreshore is the area with the most wave action. This includes the high- and low-tide areas of the beach. The offshore area extends from where the ocean bottom is shallow enough for waves to break, all the way out into the ocean depths.

You have probably heard of offshore oil drilling rigs. These are located in the deep waters off continental seashores. Figure 13-6 illustrates the different parts of a shoreline.

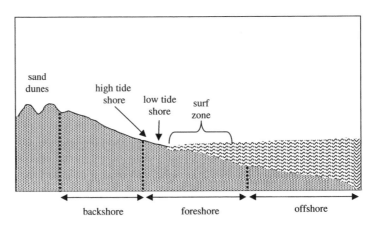

Fig. 13-6. Shorelines are divided into different land and ocean sections.

LITTORAL

The *littoral* zone is a tidal depth gradient found closest to the shore. Since there are coastal currents, onshore and offshore winds, reefs, and bays in this area, the shoreline's shape is fairly changeable. All these affecting factors make it tough to forecast water conditions accurately.

The littoral area is also where marine life, like jellyfish, is found. As many snorkellers know, there is more marine life to see near the shore than in the open ocean.

The littoral zone reaches from the shoreline to nearly 200 m out into the open ocean. It is divided into three overlapping sections: the *supralittoral* zone, the *intertidal* zone, and the *sublittoral* zone.

The *supralittoral* or *spray zone* is only washed over during very high tides or during big storms. It begins at the leading edge of the high-tide line and goes back toward dry land. The *intertidal* zone is found between the high-tide and low-tide lines. The *sublittoral* zone extends from the low-tide line out to 200 m in the water. Figure 13-7 shows the different subdivisions of the littoral zone.

Currents

Ocean waters are constantly on the move. Currents flow in complex patterns affected by wind, salinity and temperature, bottom topography, and the Earth's rotation.

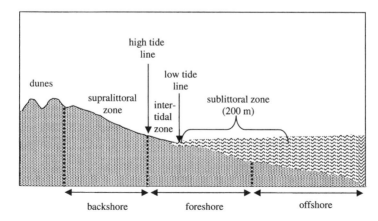

Fig. 13-7.   The ever-changing littoral zone is divided into different tide areas.

Determined by forces such as wind, tides, and gravity, currents keep the oceans stirred up. Currents move water thousands of miles. There are many different currents, but the seven main currents include the *Antarctic Circumpolar Current* (also called the *West Wind Drift*), *East Wind Drift*, *North* and *South Equatorial currents*, *Peru Current*, *Kuroshio Current*, and *Gulf Stream*. These currents move in large rotating circles called *gyres*.

Warm surface currents move from the equator to the higher latitudes, driven by atmospheric winds and the Earth's rotation. Cold surface currents start out from polar and temperate latitudes, then move south toward the equator. Like the warm surface currents, they are driven mainly by atmospheric forces. Gyres form when the main ocean currents intersect. Water flows in a circular path: clockwise in the northern hemisphere and counterclockwise in the southern hemisphere.

Large ocean currents are driven by atmospheric winds that whip the planet all year round. The winds at the equator are called the northeast and southeast trade winds. At the mid-latitudes, the winds are called the westerlies, and at the highest latitudes, the winds are called the polar easterlies. These winds blow in the same direction all year.

The Antarctic Circumpolar Current and the Kuroshio Current are two of the largest currents. The Antarctic Circumpolar Current circles eastward around Antarctica, while the Kuroshio Current, located off Japan's coast, travels up to 120 km a day at a speed of nearly 5 km/hr. A global circular current takes place when deep water is created in the North Atlantic, sinks, moves south, roams around Antarctica, and then heads northward to the Indian, Pacific, and Atlantic basins. Oceanographers estimate that it takes roughly a thousand years for water from the North Atlantic to wind its way into the North Pacific.

The *California current* is an eastern boundary current. It is wide, leisurely, cool, and shallow. Eastern boundary currents are frequently linked with upwelling.

The *Somali current*, off Africa's eastern coast, is a current with a twist. It does the impossible, it reverses direction twice a year. From May through September, the Somali current runs northward. Then from November through March it runs southward. April and October must serve as the turnaround months, like the round house for trains. As the Somali current flows northward, upwelling takes place and brings nutrients to marine life. However, when the current turns southward again, the banquet table is put away and everybody goes back to work.

Most people have heard of the Gulf Stream. It has a strong effect on the East Coast of the United States. The Gulf Stream surface current is a strong, western boundary current. It is warm, deep, swift, and fairly salty, separating open ocean water from coastal water. Western currents move commonly from tropical to temperate latitudes. The Gulf Stream commonly travels at a speed of nearly 4 knots.

The Gulf Stream is part of a greater current system that includes the *North Atlantic Current*, *Canary Current*, and the *North Equatorial Current*. From the Yucatan Peninsula in Mexico, the Gulf Stream moves north through the Straits of Florida and up the east coast of Florida. When it gets to around North Carolina of the United States, it starts to slip off into the North Atlantic toward the Grand Banks near Newfoundland.

COASTAL CURRENT

Over half of the world's population, roughly 2.7 billion, lives within 100 km of a coastline. Some countries, like Australia, have all-encompassing coastlines. Canada has the longest coastline of any country, at 90,889 km or around 15% of the world's 599,538 km of coastlines.

Most currents found along a coast are more limited. When waves hit the beach at an angle, it is called a *Longshore current*. The wave front smacks the shallow water first and then slows down. The rest of the wave folds as it comes on shore forming a current that parallels the shoreline. Larger waves hitting the beach at greater angles cause stronger Longshore currents. Longshore current action can also cause sandbars to form.

Rip currents or *rip tides* are a dangerous side effect of Longshore currents. Rip tides happen when Longshore currents, moving parallel to the beach, react seaward because of an obstacle in the bottom.

> **Rip tides** or **rip currents** are caused by a combination of Longshore currents and underwater features that react with the current.

Sandbar cuts frequently provide good spots for rip tides to occur. Swimmers should always be aware of rip tide conditions, especially on vacation or in an unfamiliar area. Rip tide warning signs are usually present, but asking about local currents is smart. Swimmers can be pulled out to sea quickly in a reacting rip current.

If caught in one of these currents, stay calm and swim parallel to the beach. When you are past the turbulent current action, you can turn and swim to shore without a problem.

UPWELLING

Offshore winds, blowing out from the land push water away from the shore. When this happens, deep, cold water rises to replace the water that has been blown out to sea. A vertical current, called *upwelling*, is then created. This forms a circular flow from the ocean bottom that brings different nutrients to the surface. Marine life increases in these nutrient-rich waters.

These nutrients come from the remains of dead organisms and fecal matter that sinks to the ocean bottom. As this material decays, nutrients are freed. They stay where they fall on the ocean's floor until an upwelling blows them back up to the surface. Large plankton increases, called *ocean blooms*, often take place after a coastal upwelling because of the nutrients that it distributes.

Coastal upwelling takes place along against the western sides of continents in the Atlantic, Indian, and Pacific. Upwelling plays an important part in the ocean ecosystem. It supports about half of the world's natural fisheries (hatching areas), although these cold waters account for only 10% of the ocean's total surface area.

What goes up must come down. *Downwelling* takes place from the opposite angle. Onshore winds, winds blowing toward the beach push water toward the land. This wind action drives the shore surface water down and outward from the beach.

Tides

We learned that water is divided into layers depending on temperature, salinity, pressure, and density. Currents and tides move seawater around changing the density. These movements change the depth of the thermocline layer. Underwater vehicles have to adjust their buoyancy to maintain a stationary depth.

When the Moon, Earth, and Sun line up, a happening called *syzygy*, shorelines experience the greatest change in high- and low-tide water levels. These *spring tides* take place twice a month, during the full and new Moon.

When the Moon is at *perigee*, or its closest distance to the Earth in its orbit, the tides are extremely high or low.

Then when the Sun and Moon form a 90° angle, like at the time of a half moon, their gravitational forces fight each other and there is a smaller change between high and low tides. These are known as *neap tides*.

There are other things that play into the equation. For example, when the Moon goes around the Earth, the planet's tilt, ocean depth, and ocean topography all affect tides. This is why not all coasts have two high and low tides a day.

Semidiurnal tides happen twice a day. The Atlantic Ocean has semidiurnal tides. It has two high tides and two low tides in one day.

Diurnal tides take place once a day. The Gulf of Mexico has only one high tide and one low tide in a 25-hr period.

Then there are the exceptions to the rule, some seas and ocean sections, including parts of the Pacific Basin, have *mixed tides*. They don't follow any set pattern, but many have one low tide followed by two high tides.

The *Bay of Fundy*, between New Brunswick and Nova Scotia holds the record for the highest tides in the world. The change in water level between high tide and low tide can be over 16 m; more than half the length of a football field.

Atlantic Ocean

There are several differences between the Atlantic and Pacific Oceans, but here are some of the highlights. The Atlantic Ocean is divided into two sections by the Mid-Atlantic Ridge; the margins of the North American and Eurasian plates. The Atlantic is also a spreading ocean except for a small subduction zone in the Caribbean Sea. The Atlantic continental margins are considered passive by geologists, since volcano and earthquake activity across the plate is rare.

The *continental shelf* in the Atlantic Ocean is a fairly flat, wide, sandy, area. It stretches gently out from the land mass of North and South America for 50 to 100 m before dropping down to the *continental slope* (4° angle) for a distance of around 1 km. The depths are between 2000 and 3000 m here.

Eventually, the continental slope begins to rise slowly. This large area, hundreds of kilometers wide, is called the *abyssal plain*. It is found at depths of 4000 to 6000 m with only the occasional *seamount* (extinct volcano) for interest. The plain rises until it comes to a hot, active volcanic gap and ridge area, the Mid-Atlantic Ridge.

The Mid-Atlantic Ridge (nearly 2000 m) is the margin of divergent plates where new seafloor is created and being pushed up from the magma below. Go back to Figure 4-5 to review how this seafloor spreading takes place along a ridge.

The other side of the ridge, closest to Europe and Africa is pretty much a mirror image of the first half as far as ocean depths and landscape is concerned.

Pacific Ocean

The first thing that most people notice about the Pacific Ocean is its size. It is huge! But because of its many subduction zones along the Ring of Fire, the Pacific Ocean is actually getting smaller. In many millions of years, barring other problems, North America and South America will be a lot closer to Asia and Australia. Who knows, by then it may be only a day trip by speed boat!

However, when people think of the Pacific Ocean, they usually think of balmy days, tropical islands, volcanic fireworks, and white beaches. That is fairly accurate, but to geologists, the Pacific Ocean is the hot spot of tectonic plate activity. It has active continental margins.

We talked about plate tectonics in Chapter 4, but in comparing the Atlantic Ocean to the Pacific Ocean, the main structural difference is probably depth. Some places in the Pacific are nearly twice as deep as the Atlantic Ocean.

For example, if you start out from the west coast of South America, the continental shelf is short, only 20–60 m wide and at its edge, the slope drops off to 8000 m (like a ball rolling off a table) into the Peru–Chile trench. On the other side of the trench, the depth rises up onto the Nazca plate around 4000 m, until it gets to the East Pacific Rise (2200 m).

On the far side of the East Pacific Rise, the Pacific Ocean floor stays around 4000 m in depth, with a number of volcanoes, seamounts, and guyots jutting up abruptly from the sea floor. Still traveling east, the Tonga Trench drops the ocean depth to nearly 11,000 m in depth. The ocean floor comes back up from this gash to 4000 m until it comes to the short continental shelf of Australia.

The Future

The oceans are critical to life on Earth. Considering rains, medicines, fish, and transported goods, the ocean plays an important role in everyone's life in some way. The oceans unite people and landmasses of the Earth. In fact, one in every six jobs in the United States is thought to be marine-related and attributed to fishing, transportation, recreation, and other industries in coastal areas. Ocean routes are important to national security and foreign trade. Military and commercial vessels travel the world on the oceans.

To acknowledge the importance of the world's oceans, the United Nations declared 1998 the *International Year of the Ocean*. This designation gave organizations and governments an important chance to increase public awareness and understanding of marine environments and environmental issues.

Quiz

1. The supralittoral zone is also called the
    (a) dune zone
    (b) spray zone
    (c) rain zone
    (d) thermocline

2. A Seamount is
    (a) a method of ocean gardening
    (b) a line of sand bars
    (c) an extinct ocean volcano
    (d) a large seahorse

3. When the sun and moon form a 90° angle, their gravitational forces fight each other and result in
    (a) zero tides
    (b) mixed tides
    (c) neap tides
    (d) perigee tides

4. The Mid-Atlantic Ridge, the margin of divergent plates, is where
    (a) nothing much ever really happens
    (b) new seafloor is created and being pushed up from the magma below
    (c) subduction is pushing the continental plate below the Atlantic plate
    (d) lots of ships sink

5. Ocean salinity measures the amount of
    (a) coral reefs formed at an atoll
    (b) salt in one millimeter of mercury
    (c) salt on your skin after swimming in the ocean
    (d) salt found in one kilogram of water

6. Diurnal tides take place
    (a) once a day
    (b) twice a day

(c) once a week

(d) once every two weeks

7. When deep, cold water rises to replace water that has been blown out to sea, it is called
   (a) unloading
   (b) upwelling
   (c) offshore current
   (d) rip tide

8. S.C.U.B.A. stands for
   (a) self-counted umbrella banking account
   (b) self-contained underwater belching apparatus
   (c) sea coral upwelling and barracuda activity
   (d) self-contained underwater breathing apparatus

9. What % of the Earth's surface is covered by oceans?
   (a) 48%
   (b) 56%
   (c) 66%
   (d) 72%

10. Which current reverses direction twice a year?
    (a) California current
    (b) Gulf Stream
    (c) Somali current
    (d) Australian current

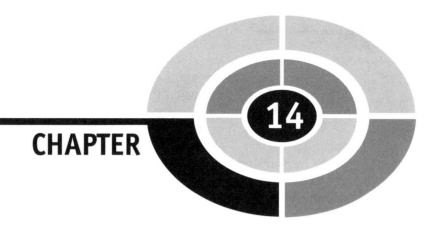

CHAPTER 14

Atmosphere

Do thoughts of space ships, hot air balloons, and kites come to mind when you think about the atmosphere? The atmosphere makes our world livable. It starts at ground level and then just goes on upward, getting thinner and colder until it finally dissolves into space. Without it, the Earth would be no more friendly than the Moon.

Many people think of the weather when they hear the word, atmosphere. Go to any party or meeting and pretty soon the weather topic comes up.

> A **meteorologist** is a person who studies the weather and its atmospheric patterns.

It can be beautiful or dreary, hot or freezing, the weather (atmosphere) shapes our lives in a big way. Along the United States Gulf Coast, the common saying is, "if you don't like the weather, then wait 10 minutes and it will change!" True of many places, the weather can change suddenly, especially around the time of seasonal shifts like spring and fall. Temperature drops of 30°F in two hours time are possible with a cold front.

What is our atmosphere made of? Air? Water? Smoke? You can't even see the atmosphere except when there is fog, rain, snow, clouds, wind, or some other atmospheric filler.

Our current atmosphere is an oxidizing atmosphere, while the primordial atmosphere was a reducing atmosphere. That early atmosphere had little, if any, oxygen. It was more a product of volcanic emissions.

Today, there are several layers that make up our protective cover called the *atmosphere*, the lower layers have oxygen that we can breathe, while upper layers have much less.

Composition

The atmosphere we all know and love contains oxygen produced almost solely by algae and plants. It is made up of a mixture of roughly 79% nitrogen (by volume), 20% oxygen, 0.036% carbon dioxide, and trace amounts of other gases.

The atmosphere is divided into layers according to the mixing of gases and their chemical properties, as well as temperature. The layer nearest the Earth is the *troposphere*, which reaches an altitude of about 8 km in the polar regions and up to 17 km around the equator. The layer above the troposphere is the *stratosphere*, which reaches an altitude of around 50 km. The *mesosphere* reaches up to 80–90 km and lies above the stratosphere. Finally, the *thermosphere*, or *ionosphere*, is still further out and eventually fades to black with outer space. There is very slight mixing of gases between layers.

The atmosphere of the Earth exists as four distinct layers. Beginning with the closest layers to the ground, they include:

- Troposphere,
- Stratosphere,
- Mesosphere, and
- Thermosphere.

These four layers are illustrated in Fig. 14-1. The location of the ozone layer is also shown.

TROPOSPHERE

The lowest of the atmospheric layers, is the troposphere, extends from the Earth's surface up to about 14 kilometers in altitude. Virtually all human activities occur in the troposphere. Mt. Everest, the tallest mountain on the planet, is only about 9 km high.

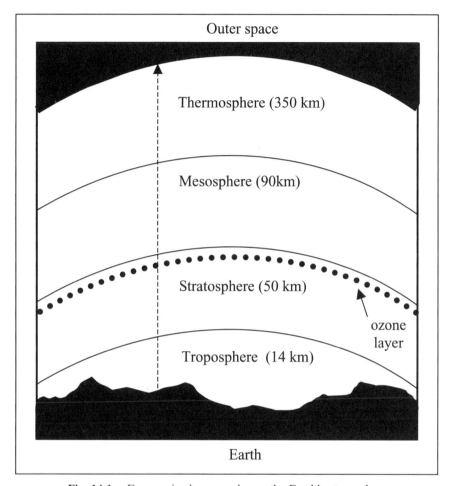

Outer space

Thermosphere (350 km)

Mesosphere (90km)

Stratosphere (50 km)

ozone
layer

Troposphere (14 km)

Earth

Fig. 14-1.   Four major layers make up the Earth's atmosphere.

Nitrogen and oxygen make up the majority of the Earth's gases, even in the higher altitudes. But it's the atmospheric layer closest to the Earth, where everything is just right to support life. At this level, living organisms are protected from the harmful cosmic radiation showers that constantly rain down on the Earth's atmosphere.

This wonder layer is called the *troposphere*. If you have ever survived a hurricane or tornado, you know it's an active place. The troposphere is the atmospheric layer where all the weather we experience takes place. Rising and falling temperatures, as well as circulating air masses keep things lively. Air pressure also adds to the mix.

When measured next to the other layers, the troposphere is fairly thin, reaching roughly 17 km up from the earth's surface.

> The **troposphere** is where all the local temperature, pressure, wind, and precipitation changes take place.

The warmest portions of the troposphere are found at the lowest altitudes. This is because the earth's surface absorbs the Sun's heat and radiates it back into the atmosphere. Commonly, as altitude increases, temperature decreases.

However, there are some exceptions. Depending on wind currents, mountain ranges can cause lower troposphere areas to have an opposite effect. Temperatures actually increase with altitude. This is called a *temperature inversion*. Generally, the temperatures at the top of the troposphere have lows of around $-57°C$. The wind speeds rise, causing the upper troposphere limits to be cold and windy. The air pressure at the top of the troposphere is only 10% of that at sea level. A mixer zone between the troposphere and stratosphere is called the *tropopause*.

STRATOSPHERE

There is a gradual change from the troposphere to the *stratosphere*, which starts at around 11 km in altitude. Here, the air flows mostly sideways. The stratosphere extends from 10 km to around 50 km. Most commercial aircraft travel takes place in the lower part of the stratosphere.

Although the temperature in the lower stratosphere is cold and constant, hovering around at $-57°C$, there are strong winds in this layer that occur as part of specific circulation patterns. However, extremely high and wispy clouds can form in the lower stratosphere. In general, there are no major weather formations that take place regularly in the stratosphere.

The stratosphere has an interesting feature from mid-level on up. Its temperature jumps up suddenly with an increase in altitude. Instead of a frosty $-57°C$, the temperature jumps up to a warm $1°C$ around 40 km in altitude in the upper stratosphere. This temperature change is due to increasing ozone concentrations which absorbs ultraviolet radiation.

The merging of the stratosphere upward into the mesosphere is called the *stratopause*.

Ozone

Since even small amounts have a significant role in protecting the life on the planet, ozone levels are important. Concentrated in a thin layer in the upper stratosphere, ozone is an exceptionally reactive form of oxygen. Atmospheric ozone is found in a layer in the stratosphere, around 15–30 km above the

Earth's surface. This ozone layer is largely responsible for absorbing most of the sun's ultraviolet (UV) radiation. Most importantly, it absorbs the fraction of ultraviolet light called *UVB*.

Ultraviolet radiation is a bad, bad thing! It causes breaks in the body's nuclear proteins leaving the door open for cancers and other health issues to take place. UVB has been connected with many serious health effects, like different kinds of skin cancer and cataracts. It is also harmful to certain crops, materials, and marine organisms.

Ozone is much less widespread than normal oxygen. The formation of the ozone layer is a tricky matter. Out of every 10 million air molecules, about 2 million are normal oxygen, but only 3 are ozone molecules. Instead of two atoms of oxygen like normal oxygen molecules (O_2), ozone (O_3) contains three oxygen atoms. Ozone has a distinctive odor and is blue in color. Regular oxygen has no odor or color.

Only through the production of atmospheric oxygen can ozone form to block the intense effects of ultraviolet radiation from reaching the surface and the plants and animals living there. In the past 30 years, there has been intense concern over decreased ozone levels. This is a big problem if we want to go on enjoying the out-of-doors and growing food in the centuries to come!

The annual "hole" or thinning of the ozone layer over Antarctica was first noticed in the spring times of the early 1980s. This area of extremely low ozone levels was having drops of over 60% during bad years. Additionally, research found that ozone depletion also took place over the latitudes of North America, Europe, Asia, Australia, South America, and much of Africa. It became obvious that ozone decreases were a global concern.

Atmospheric scientists have studied and recorded these annual and geographical fluctuations for years. There are usually cyclic downturns in ozone levels, followed by a recovery. However, as our population increases along with industrialization, global atmospheric changes are taking place as well.

MESOSPHERE

Above the stratosphere is the *mesosphere*, a middle layer separating the lower stratosphere from the inhospitable thermosphere. Extending from 80 to 90 km and with temperatures to around $-101°C$, the mesosphere is the intermediary of the Earth's atmosphere layers.

Military aircraft travel at much higher altitudes, with some classified, stealth aircraft thought to graze the boundary of the mesosphere and beyond.

THERMOSPHERE

The changeover from the mesosphere to the *thermosphere* layer begins at a height of approximately 80 km. The thermosphere is named because of the return to rising temperatures that can reach an amazing 1982°C. The different temperature ranges in the thermosphere are affected by high or low sun spot and solar flare activity. The more active the sun is, the higher the heat generated in the thermosphere.

Extreme thermosphere temperatures are a result of UV radiation absorption. This radiation enters the upper atmosphere, grabbing atoms from electrons and creating positively charged ions. Electrically charged atoms build up to form layers within the thermosphere. This ionization causes the thermosphere to also be called the *ionosphere*. Because of ionization, the lowest area of the thermosphere absorbs radio waves, while other areas reflect radio waves. Since this area decreases and disappears at night, radio waves bounce off the thermosphere. This is why far distant radio waves can often be received at night. Today, radio frequencies that can pass through the ionosphere unchanged are selected for satellite communication.

The *aurora* is found in the thermosphere. The *Aurora Borealis* and *Aurora Australis*, the northern and southern lights, are found in the thermosphere. When solar flares slam the magnetosphere and pull electrons from their atoms, they cause magnetic storms near the poles. Look back to Fig. 1-9 to review the magnetic currents that encircle the Earth.

Dazzling red and green lights are created when scattered electrons reunite with atoms, returning them to their original state. Even higher, above the auroras and the ionosphere, the gases of this final atmospheric layer begin to scatter. Several hundred kilometers above the earth, they fade into the fabric of space. NASA's Space Shuttle generally travels to altitudes between 160 and 500 km above the Earth.

JET STREAM

When watching the evening weather report, chances are good that you will hear something about the *jet stream*. This speedy current is commonly thousands of kilometers long, a few hundred kilometers wide, and only a few kilometers thick. Jet streams are usually found somewhere between 10–18 km above the earth's surface in the troposphere. They blow across a continent at speeds of 240 km/hr/hr, usually from west to east, but can dip northward and southward depending on atmospheric conditions.

> The **jet stream** is a long, narrow current of fast moving air found in the upper levels of the atmosphere.

Air temperature differences cause jet streams. The bigger the temperature differences, the stronger the pressure differences between warm and cold air. Stronger pressure differences create stronger winds. This is why jet streams fluctuate so much in speed.

During the winter months, polar and equatorial air masses form a sharp surface temperature contrast causing an intense jet stream. Stronger jet streams push farther south in the winter. However, during summer months, when the surface temperature difference is less severe, the winds of the jet stream are weaker. The jet stream blows farther north.

PRESSURE

Although air is invisible, it still has weight and takes up space. Since air molecules float freely in the vastness of the atmosphere, they get pressurized when crowded into a small volume. The downward force of gravity gives the atmosphere a pressure or a force per unit area. The Earth's atmosphere presses down on every surface with a force of 1 kg/cm^2. The force on 1000 square centimeters is nearly a ton!

> **Air pressure** is the gravimetric force applied on you by the weight of air molecules.

Weather scientists measure air pressure with a *barometer*. Barometers are used to measure air pressure at a particular site in centimeters of mercury or *millibars*. A measurement of 76 cm of mercury is equivalent to 1013.25 millibars.

Air pressure can tell us a lot about the atmosphere. If a *high pressure* system is coming, there will be cooler temperatures and sunny skies. If a *low pressure system* is moving in, then look for warmer temperatures and thunder storms.

On weather maps, changes in atmospheric pressure are shown by lines called *isobars*. An isobar is a line connecting areas of the same atmospheric pressure. It's very similar to the lines connecting equal elevations on the Earth's surface on a topographical map.

WIND

Winds are a product of atmospheric pressure. Pressure differences cause air to move. Like fluids, air flows from areas of high pressure to areas of low pressure. Meteorologists predict winds by looking at the location and strength of regional high and low pressure air masses. If the changes are slight, the day is calm. However, if the pressure differences are high and close together, then strong winds whip up.

In 1806, Admiral Sir Francis Beaufort of the British Navy came up with a way of describing wind effects on the amount of canvas carried by a fully rigged frigate. This scale, named the *Beaufort Wind Scale*, has been updated. Wind speeds are described according to their effects on nature and surface structures. Table 14-1 lists the different wind effects by increasing Beaufort numbers.

> The **wind chill factor** measures the rate of heat loss from exposed skin to that of surrounding air temperatures.

Wind chill happens when winter winds cool objects down to the temperature of the surrounding area; the stronger the wind, the faster the rate of cooling. For example, the human body is usually around 36°C in temperature, a lot higher than a cool Montana day in November. Our body's heat loss is controlled by a thin insulating layer of warm air held in place above the skin's surface by friction. If there is no wind, the layer is undisturbed and we are comfortable. But, if a sudden wind gust blows by, we get chilled. The warm protective air layer is lost and has to be reheated by the body. See Table 14-2 to get an idea of the wind chill equivalent temperatures at different wind speeds.

RELATIVE HUMIDITY

What exactly is *humidity*, anyway? Humidity is the measurement of the water vapor in air. At any set temperature, there is a maximum amount of moisture it can hold. For example, when the humidity is 75%, it means the air contains $\frac{3}{4}$ of the amount of water it can hold at that temperature. When it rains, the air is saturated, and there is 99–100% humidity.

> **Relative humidity** is the relationship between the air's temperature and the amount of water vapor it contains.

Table 14-1   The Beaufort Wind Scale gives visual clues as to a wind's speed.

Beaufort Scale #	Wind speed (km/hr)	Wind	Sign
0	< 1	calm	smoke rises vertically
1	1–3	light air	smoke drifts
2	6–11	light breeze	leaves rustle
3	12–19	gentle breeze	small twigs rustle
4	20–29	moderate breeze	small branches move
5	30–38	fresh breeze	small trees move
6	39–50	strong breeze	large branches move
7	51–61	moderate gale	whole trees move
8	62–74	fresh gale	twigs break off trees
9	75–86	strong gale	branches break
10	87–101	whole gale	some trees uprooted
11	102–119	storm	widespread damage
12	> 120	hurricane	severe damage

The air's ability to hold water is dependent on temperature. Hotter air holds more moisture. This temperature-dependent, moisture-holding capacity of air contributes to the formation of all kinds of clouds and weather patterns.

CONVECTION

In atmospheric studies, *convection* refers to vertical atmospheric movement. As the earth is heated by the sun, surfaces absorb different amounts of energy and are quickly heated.

Table 14-2   Wind chill can bring down the temperature of the body quickly.

Wind speed (km/hr)	Temperature (°Celsius)											
	−15°C	−10°C	−5°C	0°C	5°C	10°C	15°C	20°C	25°C	30°C	35°C	40°C
0	−15	−10	−5	0	5	10	15	20	25	30	35	40
5	−18	−13	−7	−2	3	9	14	19	25	31	36	41
10	−20	−14	−8	−3	2	8	13	19	25	31	37	42
30	−24	−18	−12	−6	1	7	12	18	25	32	38	43
50	−29	−21	−14	−7	0	6	12	18	25	32	38	44
70	−35	−24	−15	−8	−1	6	12	18	25	32	38	44
90	−41	−30	−19	−9	−2	5	12	18	25	32	38	45

Hot air rises. The upper floors of a house or building are warmer than lower levels. The same is true of the atmosphere. As the Earth's surface warms, it heats the overlying air, which gradually becomes less dense than the surrounding air and begins to rise.

A **thermal** is the mass of warm air that rises from the surface upward.

Thunderstorms are the result of convection. When warm air rises in some areas and sinks in others, convection is established. Thunderstorms feed on unstable atmosphere conditions.

FRONTS

Atmospheric fronts are conflicts between air masses. Depending on the air masses involved and which way the fronts move, fronts can be either, cold, warm, stationary, or occluded.

In the case of a *cold front*, a colder, denser air mass picks up the warm, moist air ahead of it. As the air rises, it cools and its moisture condenses to form clouds and rain. Since cold fronts have a steep slope face, strong uplifting is created, leading to the development of showers and severe thunderstorms. Figure 14-2 illustrates the air flow of a cold front.

In the case of a *warm front*, the warm, less dense air slips up and over the colder air ahead of the front. Like cold fronts, the air cools as it rises and its moisture condenses to produce clouds and rain. Warm fronts, with gentler slope faces, usually move more slowly than cold fronts, so the forward vertical motion along warm fronts is much more gradual. Rainfall that develops in front of a surface warm front is often constant and more wide-ranging than rain connected with a cold front. Figure 14-3 shows the gradual air flow angle and approach of a warm front.

When neither the cold front nor the warm front is pressing forward, the fronts are known as *stationary fronts*. They are at a standoff. Broad bands of clouds form on both sides of the stalemate front margin.

With an *occluded front* or *occlusion*, it is a bit of an atmospheric combination. Cold, warm, and cool airs meet forming boundaries above the ground and at the surface. When this happens, some atmospheric scientists say that a cold front has caught up with a warm front. Occluded front rainfall is a mixture of the different kinds found within cold and warm fronts.

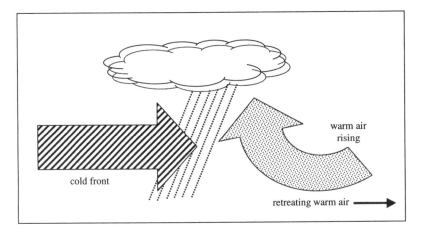

Fig. 14-2. A cold front commonly has a sharp interface with warm air.

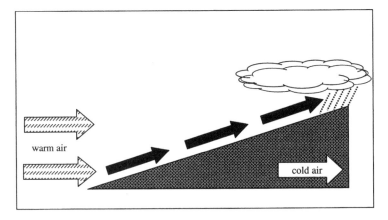

Fig. 14-3. Warm fronts are slower and more gradually approaching than cold fronts.

CLOUDS

Clouds are as varied as nature. They come in all sizes, colors, and shapes. Plus, they can change within minutes. While some clouds are happy just to be, others come with precipitation like mist, rain, sleet, hail, and snow. This makes their identification as much fun for children, as atmospheric scientists.

> A **cloud** is a combination of tiny water droplets and/or ice crystals suspended in the atmosphere.

Clouds are classified and named using Latin prefixes and suffixes to describe their appearance. For example, cloud names containing the prefix *cirr*, as in *cirrus* clouds, are found at high altitudes, while cloud names with the prefix *alto*, as in *altostratus*, are found at middle levels. Table 14-3 provides some of the common characteristics of different cloud types.

High-level clouds

High-level clouds form above 6000 m. The temperatures at these high elevations are cold, so high-level clouds are mostly made up of ice crystals. High-level clouds are generally thin and white in appearance, but can appear in a terrific variety of colors when the sun is setting on the horizon.

The most common forms of high-level clouds are thin and wispy *cirrus* clouds. Typically found at heights greater than 6000 m, cirrus clouds are formed out of ice crystals that come from frozen water droplets. Cirrus usually form in fair weather and point in the direction of the air flow at their altitude.

Cirrostratus are sheet-like, high-level clouds made of ice crystals. Although cirrostratus can blanket the sky and be many thousands of meters thick, they are fairly transparent. The sun or the moon is usually seen through cirrostratus. These high-level clouds form when a wide air layer is lifted by large-scale fronts.

Mid-level clouds

The bases of mid-level clouds usually form around 2000–6000 m. Their lower altitudes usually keep them warm enough to prevent their water droplets from freezing, but they can contain ice crystals when temperatures are cold enough.

Altocumulus may be found as parallel bands or rounded masses. Commonly a portion of an altocumulus cloud is shadowed, which helps you tell them apart from high-level cirrocumulus. Altocumulus clouds often form by convection in an unstable upper air layer. This can be caused by the gradual lifting of air before a cold front. When you see altocumulus clouds on a warm, humid summer morning, there will often be thunderstorms later that day.

Low-level clouds

Low clouds are mostly made up of water droplets, since their bases sit below 2000 m. However, when temperatures are cold enough, they can pick up ice particles and snow.

Table 14-3 There are four major cloud types at high, mid-, and low altitudes.

Cloud	Altitude (meters)	Shape	Composition
Cumulus	12,000	vertical, fluffy, defined edges and flat bases	condensed water vapor
Cumulonimbus	12,000+	massive, dark, vertical towers	water droplets and ice crystals
Cirrus	6000+	thin and wispy	ice crystals
Cirrostratus	6000	sheet-like, almost transparent	ice crystals
Altocumulus	2000–6000	parallel bands or rounded masses	high humidity and water droplets
Nimbostratus	<2000	dark, low	water or snow
Stratocumulus	<2000	light to dark gray low, lumpy masses, and rolls	weak rainfall with clear sky breaks in between
Contrail (condensation trail)	6000–12000+	long thin lines following a jet's exhaust path	water droplets freeze to ice crystals
Orographic	2000–6000+	fluffy, circling mountain peaks	condensed water vapor
Mammatus	2000–6000	light to dark gray	water droplets
Billow	2000–6000	horizontal eddies	condensed water vapor

Nimbostratus are dark, low-level clouds accompanied by light to moderately falling precipitation. However, when temperatures are cold enough, these clouds may also contain ice particles and snow.

Stratocumulus clouds generally appear as low, lumpy layered clouds that can come with weak rainfall. Stratocumulus vary in color from dark gray to light gray and may appear as rounded masses, rolls, etc., with breaks of clear sky in between.

Vertical clouds

Probably the most familiar of the basic cloud shapes is the *cumulus cloud*. Formed by either thermal convection or frontal lifting, cumulus clouds can reach up to altitudes over 15,000 m. Additionally, they free huge amounts of energy through the condensation of water vapor within the cloud itself.

Fair weather cumulus look like floating cotton balls to most people. Identified by their flat bases and fluffy outlines, fair weather cumulus show little vertical growth, with cloud tops at the limit of the rising air. Give them some frontal action, though, and fair weather cumulus become tigers, turning into gigantic cumulonimbus clouds; the citadel of violent thunderstorms.

Supplied by rising pockets of air or thermals and lifting vertically from the Earth's surface, fair weather cumulus water vapor cools and condenses to create cloud droplets. Newly formed fair weather cumulus clouds have sharply defined margins and bases, while older cumulus edges are rougher, showing cloud erosion. Evaporation around a cloud's edges cools the surrounding air, making it heavier and causing it to drop outside the cloud.

Cumulonimbus clouds are much bigger and taller than fair weather cumulus. They either build as separate soaring towers, or form a line of structures known as a *squall line*. Fed by intense updrafts, cumulonimbus clouds are the giants of the cloud forms, reaching over 15,000 m. These large, nasty clouds make up huge, towering thunderstorms called *supercells*.

Specialty clouds

Some clouds are very specialized. They form from specific events. For example, some of these clouds are formed by aircraft, earlier storms, and the presence of mountain peaks.

A *contrail*, short for *condensation trail*, is a cirrus-like trail of condensed water vapor that looks like the tail of a kite or a wide piece of yarn. Contrails are created at high altitudes when the heat from a high performance jet engine hits the extremely cold atmosphere. It condenses water vapor into a miniature cloud that forms a line behind the engine. Contrail formation depends on the atmosphere's water content and the heat characteristics of the jet engines.

Mammatus clouds, formed in sinking air, are shaped like hanging fruit. Fairly scary looking, mammatus clouds are safe and not the signal for approaching tornadoes. Actually, mammatus clouds usually appear after a thunderstorm has passed.

Air is also lifted by the land's topography. Mountains create a barrier to air currents and weather fronts. *Orographic* clouds are formed by air uplifted by the earth's shape. They are formed when rising air cools and water vapor condenses. In the United States, prevailing winds often blow from west to east, so most orographic clouds form on the western side of a mountain. Orographic clouds are also seen around mountain peaks.

Billows clouds are created from the instability connected with vertical shear air flows and weak thermal layering. These clouds, formed by winds blowing at different speeds in different air layers, are often seen at air masses margins of different densities (warm air layered over cold air). They look a lot like breaking waves.

Lenticular clouds are shaped like lenses or flying saucers. These flat clouds are found in windy areas where rising air is being cooled. After the water vapor condenses into cloud droplets, it is pulled along by the blowing wind. As the droplets warm and begin to drop, they gradually turn back into vapor and disappear. These clouds are often seen downwind of mountain ranges.

TORNADOES

Tornadoes are the children of severe thunderstorms. As speeding cold fronts smash into warm humid air, a convection of temperature and wind is formed. Winds can easily reach speeds of over 250 km/hr. Large tornadoes stir up the fastest winds ever measured on the Earth's surface. They have been measured at over 480 km/hr.

In April 1974, a huge weather system blowing cold air down from the Rocky Mountains hit rising warm, humid air from the Gulf of Mexico and caused a terrible series of storms, later referred to as the Super Outbreak. The storm stretched from Indiana where it began, to Alabama, Ohio, West

Virginia and finally Virginia. Winds were recorded over 420 km/hr spawning 127 tornadoes! Three hundred fifteen people were killed in 11 states and 6182 were injured. The Red Cross estimated that 27,500 families had suffered some extent of damage or loss in the 1,269,100 km^2 area covered by the tornadoes.

Tornadoes are usually classified into one of the following three different levels:

1. Weak tornadoes (F0/F1) make up roughly 75% of all tornadoes. They cause around 5% of all tornado deaths and last approximately 1–10 minutes with wind speeds < 180 km/hr.
2. Strong tornadoes (F2/F3) make up most of the remaining 25% of all tornadoes. They cause nearly 30% of all tornado deaths and last 20 minutes or longer with wind speeds between 180 and 330 km/hr.
3. Violent tornadoes (F4/F5) are rare and account for less than 2% of all tornadoes, but cause nearly 65% of all tornado deaths in the United States. They have been known to last for one to several hours with extreme wind speeds between 330 and 500 km/hr.

In the late 1960s, University of Chicago atmospheric scientist, T. Theodore Fujita, realized that tornado damage patterns could be predicted according to certain wind speeds. He described his observations in a table that is used today called the *Fujita Wind Damage Scale*. Table 14-4 shows the Fujita Scale with its corresponding wind speeds and surface damage.

The big problem with tornadoes is that they are unpredictable. Weather forecasters can tell when tornadoes conditions are ripe, but they don't know if or where they will strike. Think of a sleeping rattlesnake. If irritated, it will probably strike, but when and where is the question.

HURRICANES

A hurricane can exist for as much as 2–3 weeks. It begins as a series of thunderstorms over tropical ocean waters. Ocean temperature (> 26.5°C) and atmospheric pressure serve as a hurricane's fuel source. The first phase, barometric pressure drop, is called a *tropical depression*.

High humidity in the lower and middle troposphere is also needed for hurricane development. High humidity slows cloud evaporation and boosts heat release through greater rainfall. Storm intensification to a *tropical storm* marks the second phase.

Table 14-4 A tornado's strength is rated by the Fujita Wind Damage Scale.

Tornado rating	Type	Speed	Damage
F0	gale	(40–72 Mi/hr)	Light damage: some damage to chimneys, tree branches break, shallow-rooted trees tip over and sign boards damaged
F1	moderate	117–180 km/hr (73–112 Mi/hr)	Moderate damage: beginning of hurricane wind speeds, peels roofs, mobile homes moved off foundations or overturned and moving cars shoved off roads
F2	significant	181–251 km/hr (113–157 Mi/hr)	Considerable damage: roofs peeled, mobile homes smashed, boxcars pushed over, large trees snapped or uprooted and heavy cars lifted off ground and thrown
F3	severe	252–330 km/hr (158–206 Mi/hr)	Severe damage: roofs and walls torn off well-made houses, trains overturned, most trees in forest uprooted and heavy cars lifted off ground and thrown
F4	devastating	331–416 km/hr (207–260 Mi/hr)	Devastating damage: well-made houses leveled, structures blown off weak foundations and cars and other large objects thrown around
F5	incredible	417–509 km/hr (261–318 Mi/hr)	Incredible damage: strong frame houses are lifted off foundations and carried a considerable distance and disintegrated, car sized missiles fly through the air in excess of 100 meters and trees debarked
F6	inconceivable	510–606 km/hr (319–379 Mi/hr)	The maximum wind speed of tornadoes is not expected to reach the F6 wind speeds

Vertical wind shear is also a factor in a hurricane's development. When wind shear is weak, a hurricane grows vertically and condensation heat is released into the air directly above the storm. This causes even greater buildup.

> **Wind shear** describes the sudden change in the wind's direction or speed with increasing altitude.

Atmospheric pressure and wind speeds change across the diameter of a hurricane. Barometric pressure falls more and more quickly as the wind speed increases. When winds reach 62–80 knots, a storm is officially a hurricane.

Probably the one feature of a hurricane that most people look for on a weather map is the *eye of the hurricane*. The eye, roughly 20–50 km across, is found in the hurricane's center with calm or very light winds.

> The **eye of the hurricane**, the central point around which the rest of the storm rotates, is where the storm's lowest barometric pressures are found.

Just outside of the hurricane eye is the *eye wall*. This is the place where the most intense winds and rainfall are found. Wind and pressure values inside a hurricane are the same on both sides. Remember the winds are circling all the time.

Hurricane winds turn counter-clockwise about their centers in the Northern Hemisphere and clockwise in the Southern Hemisphere. Figure 14-4 shows the circular wind motion that signals a hurricane.

Hurricanes create dangerous winds, torrential rains, storm surges, flooding, riptides, and tornadoes that often result in huge amounts of property damage, deaths, and injuries in coastal locations. In 1992, Hurricane Andrew, the third strongest hurricane to ever hit the United States mainland, came ashore in Florida. It was responsible for at least 50 deaths and more than $30 billion in property damage. Table 14-5 lists the different hurricanes force categories.

Severe storms are known by different names in the different parts of the world. The name *hurricane* is used in the Atlantic and eastern Pacific Oceans, *typhoon* in the Northwestern Pacific Ocean and Philippines and *cyclone* in the Indian and South Pacific Oceans.

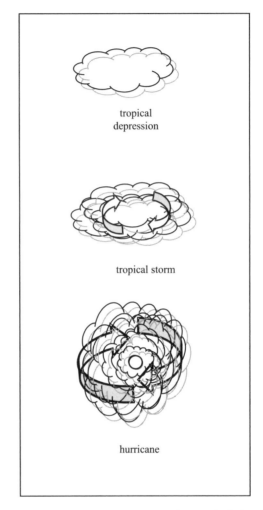

tropical
depression

tropical storm

hurricane

Fig. 14-4. Hurricanes have strong, rotating vertical cloud masses.

Since 1953, the Tropical Prediction Center has created yearly lists of hurricanes' names. As a tropical depression turns into a tropical storm, it is given the next name on the list. In alphabetical order, the names are both male and female. The name of a really destructive hurricane is never used again. Hurricanes Alicia, Andrew, Betsy, Camille, Carmen, Gilbert, Hugo, and Roxanne, to name a few, have been withdrawn from use.

Weather prediction is much more sophisticated than it was 100 years ago. Now, there is time to determine a hurricane's strength, issue warnings and list evacuation routes.

Table 14-5 Hurricanes are rated according to specific strength categories.

Hurricane category	Strength	Winds and storm surge
1	weak	65–82 knot winds, 1.2–1.7 m above normal storm surge
2	moderate	83–95 knot winds, 1.8–2.6 m above normal
3	strong	96–113 knot winds, 2.7–3.8 m above normal
4	very strong	114–135 knot winds, 3.9–5.5 m above normal
5	near total devastation	>135 knot winds, >5.5 m above normal

Quiz

1. Which of the following is a layer of the atmosphere?
    (a) angiosphere
    (b) gymnosphere
    (c) troposphere
    (d) pycnosphere

2. What is it called when a NASA T-38 jet zips through the sky and leaves a mark?
    (a) mammatus cloud
    (b) contrail
    (c) lenticular cloud
    (d) really cool

3. Which two gases make up the majority of the Earth's gases?
    (a) oxygen and methane
    (b) oxygen and propane
    (c) nitrogen and carbon dioxide
    (d) nitrogen and oxygen

4. The relationship between air temperature and the amount of water vapor it contains is known as?
    (a) indistinct humidity

(b) point source humidity

(c) relative humidity

(d) aridity

5. High-level clouds form above
    (a) 1000 m
    (b) 2000 m
    (c) 6000 m
    (d) 20,000 m

6. A long, narrow, upper atmosphere current of fast moving air is known as a
    (a) contrail
    (b) jet stream
    (c) thermophile
    (d) typhoon

7. This atmospheric layer is largely responsible for absorbing most of the sun's ultraviolet (UV) radiation
    (a) troposphere
    (b) cumulus cloud
    (c) stratonimbus
    (d) ozone

8. Conflicts between air masses are called
    (a) fossilization
    (b) fronts
    (c) precipitation
    (d) auroras

9. A sudden change in the wind's direction or speed with increasing altitude is called
    (a) wind shear
    (b) wind sock
    (c) wind chill
    (d) wind rear

10. In which layer does most of the local pressure, wind, and precipitation changes go on?
    (a) outer space
    (b) troposphere
    (c) mesosphere
    (d) stratosphere

CHAPTER

Weathering and Topography

Nothing lasts forever. This applies to winning sports teams as well as rock; the stuff of the Earth. The keys are timing and environment. Athletes do great for years, but sooner or later, they experience an injury or age catches up with them and they aren't able to maintain earlier levels of performance. This is true of any activity.

Aging and change over time are constants in nature. Whether something ages in a few months like cheese, years like wine, or centuries like monuments and old buildings, nothing stays the same forever.

Rocks and landforms are no exception. They are eventually worn away over time. As diverse as the world's environment is, so too are the many ways that rocks can be worn away. Time is the key factor. Some rock surfaces are changed over decades, while some take millions of years. Rock destruction occurs during everything from landslides to acid rain. We will look at these different rock altering mechanisms and see which ones humankind impacts, compared to those in Mother Nature's tool kit.

Denudation

Most people use the word, *weathering*, as an overall term, but it actually comes under a larger category called *denudation*. Denudation takes place when surface layers are removed from underlying rock. They are laid bare. When the wind blows constantly on a high mountain slope, there is hardly any soil left on the bare surfaces except for grains that fall into protected cracks and fissures. This is especially obvious in rock outcrops where very little soil gathers.

> **Denudation** takes place when rock disintegrates and is removed from the surface of continents.

Denudation is an umbrella word covering three main types of rock change and removal. These are *weathering*, *mass wasting*, and *erosion*. All three have the same end product, movement of rock, but they make it happen in different ways. These processes are compared below.

- *Weathering* takes place when rocks are broken down and transformed at or near the surface by atmospheric and biological agents. Weathering wears away and chemically changes rocks, but there is hardly any rock movement.
- *Mass wasting* is more active than weathering. It happens when there is a downward shift of broken rock material down slope due to gravity's pull. Mass wasting or *rock shift* can also cause loose rock to suddenly move over short distances, like during a rock slide.
- *Erosion* is a bit like mass wasting, but covers rock transport over much greater distances and is helped along by wind and/or water. Erosion causes sediments to travel over great distances, like from continental land masses to the ocean bottom.

Often, denudation takes place in a step-wise manner. First, denudation starts with weathering. Any loosened rock material is then affected by gravity, so we see the following equation that ends with mass wasting.

> Weathering + Gravity → Mass wasting

If a free-flowing stream or constant wind is added to mass wasting, then erosion takes place and becomes the product of the equation.

> Weathering + Gravity + Moving fluid → Erosion

Mass wasting and erosion, then, sometimes overlap. They can also include some type of flow, like a mudflow.

Weathering Factors

Rocks are formed at high temperatures and pressures deep within the Earth. Through tectonic activity, they can become exposed to surface conditions like low temperatures and pressures, as well as air and water. When exposed rock sees these new conditions, they react and begin to change. When this happens, it is known as *weathering*.

There are a few weathering rules-of-thumb to remember when trying to figure out the weathering pattern in a certain area. These include, rock type and structure, existing soil and weathering, slope, climate, and time. These factors, alone or combined, can make a difference in how an area of rock weathers.

Different minerals react differently to weathering action. Quartz is very resistant to chemical weathering and so are rocks containing any quartz. That is why mountains made up of rock with high amounts of quartz are still standing in areas where their sedimentary neighbors were long ago flattened.

Weathering *rinds* on basaltic stones tell geologists a lot about rates of weathering. Rinds with depths of 1–4 mm thick can point to the weathering history of a certain rock giving geologists information about surface exposure.

Texture is also an important player in weathering. A rock that is almost totally quartz, like quartzite, will be pretty open to weathering if it has a lot of joints and spaces that allow water to seep in.

Along with texture is the presence or absence of soil. The presence of soil, itself a weathering breakdown of rock, seems to add to weathering as well. When more and more surface area is exposed, either through the washing away of soil from bare rock or as rock gets increasingly jointed, then weathering can reach into more places.

Depending on composition and texture, weathering can go on at different rates in the same geographical region. This is known as *differential weathering*. It includes variable weathering of rocks with different compositions and structures, as well as a change in the intensity. That is why some cliffs with alternating hard and soft rock have curves and ridges where the harder rock has resisted weathering and the softer rock has given way.

> **Weathering** is the breakdown and disintegration of rock into smaller pieces of sediment and dissolved minerals.

An area's slope can speed or slow weathering. When a slope is steep, loosened mineral grains are washed down to the bottom with the help of gravity. They are often carried far away. Additionally, a steep, constantly exposed rock face will lose minerals as weathered grains are removed and new ones exposed. This type of slope, depending on composition of the exposed rock, may weather much faster than a less inclined slope. Gentle slopes also experience less vertical gravity pull. Loosened grains usually collect in piles, many meters thick, along the length of the slope.

Climate can also contribute to weathering changes. Heat and humidity speed up chemical and biological weathering and assist water's penetration of rock. Geologists usually find much greater weathering of limestone and marble in hot, wet climates because of the dissolving effect on calcite. Cold, dry climates have more frost-driven fracturing and weathering. We will look at these weathering factors in more detail.

Time is the biggest factor of weathering. Some hard rocks like granite are changed by different weathering factors only over thousands and millions of years. But knowing a rock's composition and its vulnerability to weathering, gives geologists important information about a region's geological history. Table 15-1 shows the different factors that affect the rate of speed of rock weathering.

Table 15-1   Weathering happens at different rates depending on different conditions.

Rock characteristics	Weathering rate		
Solubility	low	medium	high
Structure	immense	has weak points	highly fractured
Rainfall	low	medium	high
Temperature	cold	moderate	hot
Soil layer	no soil (bare rock)	thin to medium	thick
Organic activity	negligible	moderate	plentiful
Exposure time	brief	moderate	lengthy

Weathering takes place in two ways: *physical weathering* and *chemical weathering*. Physical and chemical weathering can go on at the same time. This is like a one-two punch. When a rock gets broken into pieces, more of its surface is exposed to the air. This causes more of its surface area to be uncovered to chemical weathering, which in turn breaks it down into smaller bits of rock.

Environmental factors can also cause weathering to take place at different speeds. Mechanical methods like rock smashing and cracking take place in cold and/or arid climates where water is meager. Chemical methods take place in warm and/or moist climates where there is lots of water. Biological methods take place in many different environments across the planet.

Rock Life Cycle

During weathering, rock is either worn away or shifted from one spot to another. In the course of geological time, rock goes through an entire life cycle or *rock cycle*. One rock's lifetime might include being blasted out of a volcanic vent to the surface, settling back on the earth as a layer of volcanic ash, being lithified with other sediments into a sedimentary rock layer, then being pushed down at a subduction zone to be transformed by pressure to metamorphic rock. Should the metamorphosed rock come in contact with a magma chamber or hot spot, it might melt and be sent back to the surface. The cycle would be complete or it would start all over again!

A rock can go through many changes and forms over billions of years. Figure 15-1 gives you an idea of a rock's life cycle.

For more information about igneous, sedimentary, and metamorphic rock types, refer back to Chapters 6–8 for specific characteristics. Otherwise, just remember that given enough time, a particular rock can have an interesting and active history.

Like people, a rock is affected by its environment. We have learned how different rocks form, now let's look closer at the factors that play a part in a rock's aging.

PHYSICAL WEATHERING

The activities of physical and mechanical weathering create cracks in rock that act as channels for air and water to get deeper into a rock's interior.

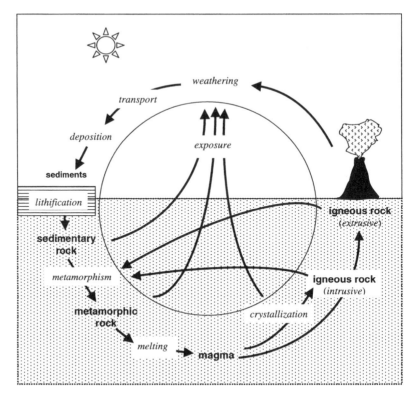

Fig. 15-1. The rock cycle shows the many different paths a rock can take over time.

During weathering, rock is constantly being broken into smaller pieces and the surface area that is exposed to air and water gets larger. Both of these actions add to the overall chemical weathering.

> **Physical weathering** happens when rock gets broken (cracked, crumbled, or smashed) into smaller pieces without any change to its chemical composition.

Physical or *mechanical weathering* is the breakdown of large rocks into smaller bits that have the same chemical and mineralogical make up as the original rock. Everyone is familiar with the breakdown of rock into smaller and smaller portions:

boulders ⇨ pebbles ⇨ sand ⇨ silt ⇨ dust

This size-graded breakdown takes place in different ways.

JOINTS

Many rocks are not solid all the way through. They have a lot of different size cracks and fractures caused by stress, called *joints*.

Most jointing takes place in rock as a result of the activities of internal or external forces. These include:

(a)   Cooling and shrinking of molten matter;
(b)   Flattening and tightening of drying sedimentary strata; and
(c)   Plate tectonics.

Joints are often arranged in sets of vertical parallel lines. The largest, most visible fractures are called *master joints*. Sometimes, multiple sets of joints intersect at nearly right angles and create a joint system made up of crossing vertical and horizontal joints.

In sedimentary rocks, various sets of joints often match up with planes separating strata. You see the flaking away between layers of mica as stratification planes allow. Along with joints, rocks can also fracture between individual crystals or grains. Weathering factors, like water, enter tiny gaps between the grains, leading to grain-by-grain disintegration of rocks.

The two most important forms of physical rock breakdown are *joint block separation* and *granular disintegration*. Break up of rock happens in several different ways. Some of these include:

- Frost
- Salt-crystal growth
- Unloading (weight)
- Expansion and contraction due to change of temperature and wetting/drying cycles
- Biological weathering, and
- Chemical weathering.

FROST

Frost wedging is a form of mechanical weathering (that is, weathering that involves physical, rather than chemical change). Frost wedging is caused by the repeated freeze–thaw cycle of water in extreme climates. Freeze and thaw weathering is common when temperatures drop below freezing at night and rise during the day.

Most rocks have small cracks in them, called *joints*. When it rains, rainwater collects in these joints. As the day cools and temperatures drop at night below freezing, the water inside the joints freezes.

As water freezes into ice, it expands. You have probably seen this for yourself. Have you ever put a carbonated drink in the freezer to cool it quickly and forgotten about it? When you come back a few hours or days later, the aluminum can is stretched and deformed from the pressure of the expanding liquid. The same thing happens when you fill a container with water, screw on the lid, and then put it into the freezer. You have to leave a little room at the top or the container will burst. If it has a loose fitting lid, the lid will be lifted right off as the ice expands up and out.

Within rock fissures, expanding ice puts pressure on the joints in the rock. When the pressure gets too great, the joint expands. Sometimes, the rock will split the first time. Or after many repeated freezes and thaws, the joints can't take it anymore and finally crack.

During warmer summer months, some ice melts back to free-flowing water. This seeps into joints and is frozen back into ice at night. This further wedges joints apart.

> **Frost wedging** happens when rock is pushed apart by the alternate freezing and thawing of water in cracks.

Frost action is best seen in wet climates with many freezing and thawing (arctic tundra, mountain peaks) cycles. Frost splits rocks into blocks by *joint block separation* and also wears away the edges of blocks grain-by-grain, rounding the surfaces.

In areas with a lot of glaciers, rock ridges experience a good deal of frost wedging. *Arêtes*, the steep ridges between glaciated valleys, are commonly sharpened through the frost wedging process.

SALT WEDGING

When you think of deserts, what do you think of? (Cactus, sand, and lizards, right?) Most people think of deserts as barren, dry, windy, inhospitable landscapes with only circling vultures to keep you company. The one thing deserts don't have is an abundance of water. Everything in a desert is *desiccated* (dried out) much of the time. This is a harsh environment for man and beast! It is also a good place for weathering to take place.

Salt wedging, caused by the growth of salt crystals, is an important rock breaking force in the desert. In dry and semi-desert environments surface water and soil moisture evaporates quickly. When this happens, salts dissolved in the water fall out of solution and crystallize.

Growing salt crystals, such as halite (NaCl), calcite ($CaCO_4$), and gypsum ($CaSO_4$), exert pressure on the rock. Over time, this force wears bedrock away grain by grain.

Salt wedging is especially active at places where ground water is discharged and then evaporates. Ground water that is discharged at the base of a sandstone cliff, gives rise to particularly active salt crystal growth.

Niches enlarged in this way have been used as homes by native peoples. Cliff villages of native Americans in Arizona and New Mexico have been preserved at the bases of sandstone cliffs in the southwestern United States.

UNLOADING

We have seen how rocks that are buried within the crust are under pressure from the weight of overlying rock layers. Over time, erosion and mass wasting takes away upper rock layers and the pressure on the lower rock layers is lifted. When this happens, the internal rock volume expands and the outer rock layers are cracked away. This shedding of outer rock layers due to internal pressure changes is called *unloading*. It's like a snake that sheds its skin as it grows larger.

> **Unloading** happens when there is a release of internal rock pressure from erosion and the outer layers of a rock are shed.

When rock on the Earth's surface is worn away, it is called *exfoliation*. Exfoliation can take place as large, flat, or curved sheets of rock shear away from the primary rock. When this happens, a mass of igneous rock may be exposed which expands upwards forming a dome. Then, it expands further and breaks into sheets that are parallel to the surface.

An awesome example of an exfoliation dome can be found in Yosemite Valley, California in the United States, where thousands of visitors go each year to view the majestic Half Dome.

EXPANSION AND CONTRACTION

Simple ups and downs of daily temperature can cause microscopic changes in a rock's volume by expanding and contracting it with heat. Over thousands of years, stress created by such temperature swings add up and crack the rock. The constant stress basically wears it out, like when you bend a paper clip back and forth in the same spot over and over, until the stress causes the metal to wear out (fatigue) and break.

Wetting and drying cycles work in a related way. In sedimentary rock, baking and wetting of fine sediments causes cracking. You have probably seen desert basins or stream beds with mud that has dried and broken into hexagonal pieces of anywhere from 2 to 40 cm across. When these dried pieces get hard enough, they look like flattened, angular bowls that you can pick up and examine.

BIOLOGICAL WEATHERING

Biological weathering is a blend of both physical and chemical weathering. Tree and plant roots grow into rock fissures to reach collected soil and moisture. As they grow, roots get thicker and push deeper into the crack. Eventually, this constant pressure cracks the rock. Then, the more the rock is fractured, the easier it becomes to crack. Figure 15-2 shows how roots can cause rocks to fracture further by widening existing cracks. Both root growth and burrowing in rock gaps can cause rock fractures to open wider and become more exposed to future weathering.

Roots growing at an angle can also push rock aside as they get larger and more embedded. You have probably seen this in your own neighborhood where tree roots crack and force up sidewalks.

In biological weathering, there is also an element and nutrient exchange. Bacteria and algae growing in cracks and on rock surfaces have acidic

Fig. 15-2.   Biological weathering takes place when roots push their way into rock joints and fissures.

processes, which add to the chemical weathering. Plants get needed minerals from rock and upper soil layers. This is described in greater detail below when we look at how soil horizons are formed.

CHEMICAL WEATHERING

As mechanical weathering breaks rock apart, the total area of an exposed rock surface increases and *chemical weathering* increases. Chemical weathering breaks down rock through reactions with a rock's elements and mineral combinations. These then change the rock's chemical structure and form. It is much more subtle than physical weathering.

Chemical weathering has been going on for millions of years, but with the addition of man-made industrial pollutants into the Earth's air and water, some forms of chemical weathering have increased.

> When rock and its component minerals are broken down or altered by chemical change, it is known as **chemical weathering.**

Chemical weathering takes place in one of the following ways:

- *Oxidation* = reaction with O_2
- *Hydrolysis* = reaction with H_2O
- *Acid action* = reaction with acid substances (H_2CO_3, H_2CO_4, H_2SO_3)

The most important natural acid is *carbonic acid* which is formed when carbon dioxide dissolves in water ($CO_2 + H_2O \leftrightarrow H_2CO_3$). Carbonate sedimentary rocks, like limestone and marble, are extra sensitive to this type of chemical weathering. Gouges and grooves that are often seen on carbonate rock outcrops are examples of chemical weathering.

There are several chemical reactions that occur to speed along chemical weathering. Probably the best known and most newsworthy chemical weathering is that of *acid rain*. Acid rain is formed when chemicals in the atmosphere react with water and return to the earth in an acidic form as rain. When this happens, the process is known as *dissolution*. When acid rain falls on limestone statues, monuments, and gravestones, it can dissolve, discolor, and/or disfigure the surface by reaction with the rock's elements. Historical treasures such as statues and buildings, hundreds to thousands of years old suffer from this kind of weathering; a by-product of industrial pollution.

Chemical weathering replaces original rock minerals with new minerals. These replaced minerals may have very different mechanical characteristics such as strength and malleability. For example, if clays are formed, they may not be as rigid as the existing rocks might have been, but are much more pliable. Chemical weathering constantly weakens rocks by increasing the chances of *mass wasting*.

- **Oxidation** takes place when oxygen anions react with mineral cations to break down and form oxides, such as iron oxide (Fe_2O_3) and soften the original element.
- **Solubility** is the ability of a mineral to dissolve in water. Some minerals dissolve easily in pure water. Others are even more soluble in acidic water. Rain water that combines with carbon dioxide to form carbonic acid ($H_2O + CO_2 = H_2CO_3$) becomes naturally acidic.
- **Hydrolysis** takes place when a water molecule and a mineral react together to create a new mineral. The transformation of the feldspar mineral, orthoclase, into clay is an example of hydrolysis.
- **Dissolution** goes on when acids in the environment like carbonic acid (water), humic acid (soil), and sulfuric acid (acid rain) react with and dissolve mineral anions and cations.

In chemical weathering, water nearly always has a part. Carbon dioxide dissolves in rainwater forming carbonic acid, which dissolves limestone rock and carries it away in solution as calcium carbonate. When limestone rock is dissolved away over long periods of time by underground streams, intricate caves and channels are formed.

Chemical weathering, faster for limestone than sandstone, is speeded up by heat. It takes place the quickest at the sharp, thin edges of rocks with larger surface areas, but less volume. Because of this, they erode faster. The same thing happens with limestone buildings. Greek ruins, over 2500 years old, show pitted, etched limestone columns, edges and surfaces that have been weathered by chemical dissolution.

SPHEROIDAL WEATHERING

When rock fragments, made sharp and angular by fracturing, are slowly smoothed at the corners and worn down, it is a type of mechanical weathering. On an even smaller scale, grains of sand and pebbles in nature form from this type of weathering.

Think of a child's rock tumbler that weathers water and rock together mechanically. These simple cylinders that spin small rocks round and round for days and weeks, non-stop, smooth rocks by weathering them. This

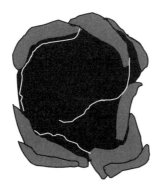

Fig. 15-3. Spheroidal weathering flakes away curved layers of rock.

time-related activity then transforms small, rough rocks into smooth, shiny "gems."

On a larger scale, boulders sometimes have a type of fracturing that cracks away from the central "core" of the rock like layers of an onion. The pieces that break off are rounded and copy the shape of the parent rock. When this happens, it is called *spheroidal weathering*. Figure 15-3 gives you an idea of the rounded look of weathered rock layers.

Spheroidal weathering is usually found during the weathering of homogeneous rock. Although rounded, cast off layers are formed by weathering, geologists have not completely figured out the sequence or timing of spheroidal weathering.

Soil Erosion

Erosion converts soil into sediment. Chemical weathering produces clays on which vegetation can grow. A mixture of dead vegetation and clay creates soil which contains necessary minerals that plants need for growth.

> **Soil** exists as a layer of broken, unconsolidated rock fragments created over hard, bedrock surfaces by weathering action.

Most geologists talk about soil as being part of three layers called *soil horizons* or *soil zones*. These three soil horizons are commonly recognized as "A," "B," and "C," but it is important to remember that not all three horizons are found in all soils. Figure 15-4 illustrates the way soil horizons are stacked on top of each other.

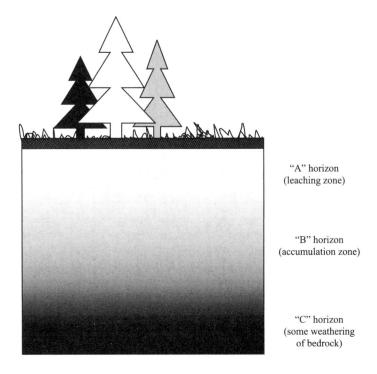

"A" horizon
(leaching zone)

"B" horizon
(accumulation zone)

"C" horizon
(some weathering
of bedrock)

Fig. 15-4.   Soil horizons form in a gradient with existing rock at the bottom.

Soil horizons are described from the top soil layer down to the lowest soil and bedrock level and are as follows:

- **"A" horizon** includes the surface horizon, a zone of leaching and oxidation, where penetrating rain water dissolves minerals and carries the ions to deeper horizons. It also holds the greatest amount of organic matter.
- **"B" horizon** describes the middle horizon, a zone of accumulation, where ions carried down by infiltrating rain water are reconnected to create new minerals. Blocky in texture it is made up of weathered rock mixed with clay, iron, and/or aluminum.
- **"C" horizon** includes the bottom horizon, a zone of unconsolidated, weathered original rock.

SOIL TYPES

Just as there are three different soil horizons, there are also several factors that determine which type of soil will form. These include structure, rainfall (lots or little), solubility, temperature (hot or cold), slope (gentle or steep), vegetation (types and amount), and weathering time (short or long.) Singly or in combination, soils form as a result of many different factors. A key factor in naming major soil types is rainfall amount. Everyone from toddlers making mud pies to petroleum geologists looking for oil can tell whether a soil is wet or dry, hard or soft.

Geologists have named three basic soil types based primarily on water content. These are the *pedocal*, *pedalfer*, and *laterite*.

The *pedocal* is found in dry or semi-arid climates with little organic matter, little to no leaching of minerals and is high in lime. Most nutrient ions are still present. In places where water evaporates and calcite precipitates in the "B" horizon, a hard layer called the *caliche* or *hardpan*, is formed. Pedocal soil also collects in areas of low temperature and rainfall and supports mostly prairie plant growth.

Pedalfer soil is found in wetter environments and contains greater amounts of organic matter and leaching. It is enriched with aluminum and iron after many other soluble nutrients are leached out. This type of soil is found in areas of high temperatures and humid climates with a lot of forest cover.

Laterite, the soggiest type of soil, is found in tropical and sub-tropical climates and is high in organic matter. Because of high equatorial rainfall in very wet climates, there is widespread leaching of silica and all soluble nutrients. Iron and aluminum hydroxides are left behind and cause well-drained laterite soils to be red in color. Besides iron and aluminum ores, laterites can also form manganese or nickel ores.

REGOLITH AND SLOPES

Remember from Chapter 7, that *regolith* is a collection of many different soil and rock types. It's the loose rock matter, like volcanic ash and wind-driven deposits that are scattered around on bedrock.

All rock surfaces, except for the super steep and the fairly new (geological time) are covered by a layer of weathered material. Generally, the growth of plants in a specific area contributes to soil development by holding it in one place. This allows the soil that builds up on a rock to take place in a layered way.

At the deepest level is bedrock, with different types of overlying regolith and finally a scattering of soil on top. This is shown simply in the series below.

bedrock ⇨ regolith ⇨ residual regolith ⇨ transported regolith ⇨ soil

Table 15-2   Bedrock and regolith layers are topped with a thin cover of soil.

Rock type	Characteristics
Soil	top layer of regolith (1–3 m) mixed with organic matter
Transported regolith	sediment transported and laid down by erosion (water current, waves, wind, ice) and mass wasting
Residual regolith	weathered rock from settled lower bedrock
Regolith	layer of weathered rock
Bedrock	solid unchanged rock

Of course, depending on the region, this soil progression might not all be visible. But just knowing how it all stacks up can give you an idea of what is missing in any one rock formation. Table 15-2 describes what each layer in the soil development column contains.

MASS WASTING

In general, *mass wasting* doesn't happen with a lot of flow-type movement. It moves material a fairly short distance, compared to the longer distances that sediments are carried by rivers.

> **Mass wasting** describes the slow or sudden movement of rock downslope as a result of gravity.

Mass wasting has several factors that affect it. These include gravity, types of soil and rock, physical properties, types of motion, amount of water involved, and the speed of movement.

Gravity is the main influence on mass wasting. It is always pulling things down. When rocks are piled on a steep mountain slope, there is a high amount of friction that holds them to the slope. However, gravity is pulling the rock downward. The downslope pull of gravity that causes mass wasting is known as *shearing stress*. The steeper the slope, the greater the shearing stress. Figure 15-5 illustrates the difference of slope angle on shearing stress.

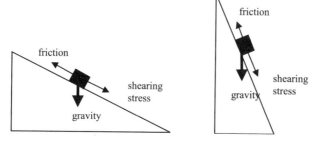

Fig. 15-5.   When a mass is the same on two slopes, the greater the slope angle, the greater the shearing stress.

> **Shearing stress** is linked to the mass of the rock being pulled downward by gravity and to the slope's angle.

The counteracting force that works against shearing stress is friction or with a large body of rock, it is called *shear strength*. When the amount of shear stress is higher than the shear strength, something has got to give. A quick movement like an earthquake acts as a trigger. It provides just enough energy to overcome the last bit of friction and allow gravity to pull everything down.

There are several different mass wasting rock movements.

ROCK FALLS AND SLIDES

Rock slides and falls with little to no loose flowing material. Think of it more like huge, shifting boulders than a bunch of loose, flowing mud and pebbles. Giant boulders or sections of earth shift position because of shear stress and move as a whole down a slope.

> A **rockfall** takes place when large amounts of rock free fall or shift downward from very steep areas of a mountain slope.

Generally, a *rockfall* happens along the sides of a steep mountain, with little plant life, but can also involve cliffs, caves, or arches. Small rockfalls are fairly common, but huge rockfalls are rare because of the amount of force needed to move tons of rock. Small rockfalls take place mostly because of weathering, while earthquakes use their energy to cause large and small rockfalls.

Talus is the pile or cone-shaped mound of broken, blocks of rock that gather at the foot of a cliff or steep mountain edge from rockfalls. When there are massive amounts of broken rock below a deep scar, geologists know that this is a result of a major sudden rockfall or many smaller rockfalls. Figure 15-6 shows how talus collects at the base of a cliff. A good example of this type of large, blocky talus pile can be seen at Devil's Postpile National Monument in the western United States in California.

Rock and soil can experience a *rockslide* or *landslide* when the moving rock slips on an underlying layer, like sedimentary rock, and moves in a solid sheet downward. Again, there is little loose flow.

> A **rock slide** takes place when a fairly solid chunk of rock or soil slides downward along a clearly visible surface or plane.

Slides take place because of the buildup of: (1) internal stress along fractures; (2) undercutting of clay layers and slopes by water (rivers and glaciers); and (3) earthquakes. Rock slides form deep, wide scars and a massive pile of highly broken debris (talus). Although rockslides are fairly rare, when they do happen, they can move at speeds of up to 150 km/hr.

When an area of rock or soil moves a fairly short distance, it is sometimes called a *slump*. A slump of soil can rotate a bit along a slippage plane.

Fig. 15-6. A talus pile builds up at the base of a cliff from rockfall rubble.

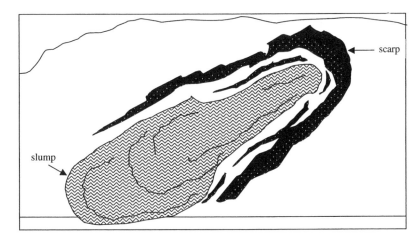

Fig. 15-7. When rock slumps down a valley, it forms a scarp where it breaks away.

Although the plants and soil on top may not move much, the entire area can turn slightly as it slips down the slope. When a section of soil breaks away and slumps downward from a cliff, the original site of the break is called a *scarp*. Figure 15-7 shows an area of slump below a scarp.

ROCK FLOW

When rock flow takes place, there is usually some amount of water involved. This includes soil flow, snow avalanches, and pyroclastic flows from volcanic eruptions that we saw in Chapter 11. However, any disorganized flow is considered a *rock flow*.

When soil is the main flowing substance, it is called an *earthflow* and involves less water than mudflows. Earthflows happen along low-angle, less stable parts of a cracked slope that contains some amount of clay in its makeup. These are not too dangerous because they are slow, but earthflows can destroy property.

When the ground becomes saturated with water, it's called a *mudflow*. Mudflows are often seen in deserts and dry environments during heavy rains following a dry period, when the soil becomes saturated. After a mudflow, a layer of debris is smeared across the valley floor. It can happen in minutes or an hour. It's dangerous because of its speed and the fact that it carries mud, rock, boulders, and whatever else is flowing (like buildings) down valleys or steep foot hills. Mud flows usually follow an old stream route. When completely soaked, everything flows together in a single, swift gush of water, soil, rocks, plants, and trees. This is known as a *debris avalanche*.

People who live in fairly dry climates like southern California make the mistake of building homes and businesses on hills or cliffs overlooking the ocean or some scenic view. This is fine as long as the weather stays dry, but weeks of rain can cause mud slides that undercut the building and cause supporting soil to slide down the hill. Besides being dangerous and costly, this hazard could be easily avoided if houses were only built on solid rock.

The slow flowing movement of soggy soil from higher ground to lower ground is called *solifluction*. Soil flow can take place in the thawed soil layer covering permafrost. It is limited to tundra beyond the tree line and is affected by the summer sun's heat. When the top soil layer melts and begins moving over the solid permafrost below, then there is solifluction.

Soil *creep* is the most common mass wasting process; it takes place on almost all land surfaces and slopes. The main difference between other types of movement and creep is rate of movement. Creep is extremely slow. It causes the land to be deformed from its original shape, but is too slow to cause shear problems like in a landslide.

> **Soil creep** takes place as a very slow (less than 1 cm/year) downhill movement of soil and regolith.

Soil creep can be seen most easily on steep, bare, wet slopes. Depending on how damp the soil is, creep can move along as fast as 5 cm/year. A lot of different conditions can cause creep. Some of these include: drying and wetting, heating and cooling, freezing and thawing, walking and burrowing by animals, and shaking by earthquakes.

When you can see sedimentary layers curving gently downhill along with tilted fences, poles, and walls, it can be caused by soil creep.

Topography

Unlike what early people believed, the Earth is far from flat. Even in plains or deserts where you can see for miles, there are slight changes in the *elevation* of the land. A mountain like Pike's Peak in the Colorado Rocky Mountain Range rises 4301 m above sea level.

> In geology, **elevation** is the height (altitude) of an area or landform above sea level.

In fact, spectacular changes in altitude, give the land masses their personality. They make Earth Science such a diverse field. One story tells of how the words of the song, "America the Beautiful" were written by Katharine Lee Bates, an English professor from Wellesley College, Massachusetts, in July of 1893. Professor Bates, who was teaching a summer class at Colorado College, took a carriage and wagon ride to the top of Pike's Peak with a group of students and other faculty members. After seeing the magnificent summit view of other mountains and far away plains, Bates was inspired to write the words to the song upon returning to Colorado Springs that evening. "America the Beautiful" was first published in *The Congregationalist*, a weekly journal, on July 4, 1895. The poem was published with music (composed by Samuel A. Ward) in 1910.

Besides singing about the land, we also use maps to describe the landscape. Most maps give geological information as well as directions. If you study a road map that lists the highest elevations in Colorado, the elevation of Pike's Peak would be listed as 4301 m. The map will also show the heights of many of the other 54 mountains in Colorado with heights over 4200 m. Mount Ebert, the highest mountain in the state, would be listed at an elevation of 4400 m.

Topography is the recording of the contours and physical features of an area, especially on the land.

The word *topography* comes from the Greek words *topos* meaning place and *graphein* meaning to write. Geologists use topographical maps to give better information about the contours or highs and lows and curves of a place. There are thousands, if not millions, of topographical maps that show the height and shape of different areas in the United States. Many more maps exist for land masses around the world.

Most people use maps to get directions or to just learn the "lay of the land." Today, we have the Internet for maps and driving directions. Some vehicles have GPS (global positioning system) to direct them to their destinations.

Figure 15-8 shows a simple picture of an area and its topographical map. Topographical maps include the elevation in feet or meters of an entire area.

Two-dimensional maps, with only the length and width of an area shown, are called *base maps*. These are the standard road maps that get people from here to there. These maps are drawn in relation to the north and south poles, generally with north at the top.

Two-dimensional maps include *longitude* and *latitude* lines to increase the accuracy of pin-pointing a specific location. Longitude lines run north and south between the poles (‖), while latitude lines run horizontally (=) around

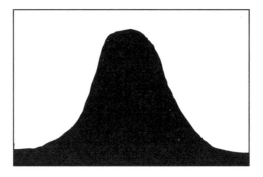

visual view of mountain

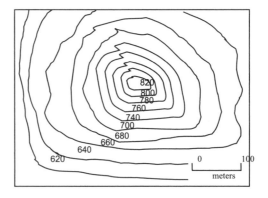

topographical map
(*contour interval 20 meters*)

Fig. 15-8.   A mountain as it appears on a topographical map with elevations.

the globe. Site locations are often given by their nearness to a longitude and latitude intersection.

Scale, Relief, and Contour

The art and science of expressing the physical features of the Earth's surface is called *cartography*. A map's accuracy is determined to a large degree by the accuracy with which distance is represented. Scale is crucial to understanding any map. Depending on a map's scale, a trip can take an hour or a day. If a map scale is listed as 1 cm/km compared to 1 cm/10,000 km, a trip would be totally different with different requirements. In order to stay accurate, road maps in growing cities must be constantly updated with new streets, housing, parks, and other highlights added as they are developed.

Geological maps are updated less often, unless some new development has occurred, like a volcanic eruption. Remember after the Mount St. Helens eruption, the mountain peak was reduced by 400.5 meters. This new elevation of the mountain's peak had to be changed on topographical maps of the area.

> **Scale** represents the proportion between a map's unit of distance and a unit of distance on the Earth's surface (for example 1:1000 for 1 cm to 1 thousand meters).

Many maps drawn by the United States Geological Survey often use a map scale of 1 inch roughly equal to 1 mile.

Geologists get most information from three-dimensional topographical maps. A three-dimensional map is more complete since it shows height, as well as length and width. When a map shows height or *elevation* it gives the reader an idea of the height differences between the highest places in an area (mountains) and the lowest places (valleys) in addition to their location.

> The difference between the highs and the lows of a land's surface is called a topographical map's **relief**.

An Earth globe is a map of the entire Earth plotted as a sphere. Some globes show relief by providing mountains, ocean ridges, and other elevation changes as raised surface features. So when you run your hand over the surface, you can easily feel the height differences across different continents.

Contour or *contour lines*, give you an idea of the shape, as well as the elevation increases of an area. If you walk at the same elevation around a single landform, like a mountain or a hill, you wouldn't go uphill or downhill, but travel in a circle, always at the same level. You might climb over boulders, cross streams, and weave through woods, but to stay on one contour line, you have to stay at the same elevation.

> A **contour line** passes through all the points of the same elevation (altitude) above sea level or at a specific distance from a land form.

Look at Fig. 15-8 again. The mountain's contour lines in the figure are shown at 20-m intervals. They provide the rate of elevation rise all the way around the mountain at known intervals. Contour lines also give an idea of the mountain's physical outline from a "bird's eye" perspective.

> The **contour interval** records the vertical distance between two neighboring contour lines.

Reading a topographical map is easy if you remember a few rules-of-thumb that map makers use when drawing the contours of a certain area. These rules are as follows:

- All points on a contour line have the same vertical elevation (altitude);
- Contours have a set distance used between every contour interval (like 20 m or 1000 feet);
- A contour separates all points of higher elevation from those of lower elevation;
- Contours do not intersect or separate (they can join at vertical cliffs);
- Every few contour lines (like 4 or 5), the line is drawn in bold to make reading the map easier;
- Tightly drawn contours show steep slopes, while widely spaced contours show gradual slopes; and
- Dips of lower elevation have *hatch* marks (///) to set them apart from raised hills.

Weathering and topography have a big effect on sediment transport and chemical breakdown. The steeper and more rugged the land, the greater the pull of gravity works along with wind, water, and glacier ice to move sediments. When sediments move quickly, there is less chemical breakdown, but more physical and mechanical rock fracturing. Remember the rounding of sand and stones at the beach in the turbulent surf?

Where land is flat, there is much less sediment transported except for that done by wind and stream movement. Also, in gently sloping areas, physical rock breakdown is much less, while the chemical breakdown of minerals is higher.

The Earth is ancient and new, solid and molten, windy and calm, building up and wearing away. It is the most diverse and beautiful planet in our solar system with mysteries yet undiscovered. It is our home.

Quiz

1. Mass wasting is a combination of
    (a) weathering plus centrifugal force
    (b) gravity plus magma

(c) weathering plus gravity
(d) erosion plus pyroclastic flow

2. Frost wedging is caused by
    (a) Jack Frost
    (b) the repeated freeze–thaw cycle of water in extreme climates
    (c) mudflows
    (d) lava backup

3. The proportion between a map's unit of distance and a unit of distance on the Earth's surface is called
    (a) geography
    (b) a road trip
    (c) an eye chart
    (d) scale

4. The soil found in dry or semi-arid climates with little organic matter and little to no leaching of minerals is called
    (a) pedocal
    (b) pedalfer
    (c) laterite
    (d) muck

5. A contour on a topographical map is the
    (a) line passing through all points with the same altitude above sea level
    (b) line that always marks the best restaurants in an area
    (c) one point that marks the distance between the mountain summit and the valley
    (d) one point written in pounds per square inch

6. When rock disintegrates and is removed from the surface of continents, it is called
    (a) an eruption
    (b) denudation
    (c) shedding
    (d) lithification

7. Which weathering type covers large distances and uses wind and water?
    (a) spheroidal weathering
    (b) sculpture
    (c) slippage
    (d) erosion

8. Chemical weathering happens in all but one of the following ways
   (a) acid action
   (b) hydrolysis
   (c) intoxication
   (d) oxidation

9. Most rock jointing takes place as a result of all, but which of the following?
   (a) cooling and shrinking of molten matter
   (b) peer pressure
   (c) flattening and tightening of drying sedimentary strata
   (d) plate tectonics

10. A type of fracturing that cracks away from the center of the rock like layers of an onion is known as
    (a) spheroidal weathering
    (b) biological weathering
    (c) solubility
    (d) onionization

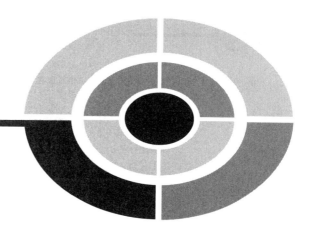

Part Three Test

1. Volcanoes are classed into 3 types
   (a) pretty hot, sweaty hot, burning hot
   (b) plutonian, volcanian, and strombolian
   (c) active, dormant, or extinct
   (d) watch your step, could you move a little faster and run

2. The slow or sudden movement of rock downslope as a result of gravity is called
   (a) intrusion
   (b) topography
   (c) extrusion
   (d) mass wasting

3. Lava lakes, a product of Hawaiian type eruptions, are
   (a) established when people don't have central heat
   (b) formed when fluid basalt ponds in vents, craters, and low spots
   (c) the opposite of Alpine lakes
   (d) never seen

4. The art and science of expressing the physical features of the Earth's surface is called
   (a) cartography
   (b) photography
   (c) biology
   (d) iconography

5. The repeated freeze–thaw cycle of water in extreme climates causes
   (a) lava flows
   (b) increased plant growth
   (c) frost wedging
   (d) colds and flu

6. The Latin word *vulcanus* means
   (a) white flower
   (b) fire breathing
   (c) dinner is ready
   (d) first

7. What method is used by whales and dolphins to find out what is happening in the ocean around them?
   (a) photosynthesis
   (b) temperature sensing
   (c) echolocation
   (d) circumnavigation

8. Tides that take place once a day are called
   (a) annual tides
   (b) diurnal tides
   (c) nocturnal tides
   (d) semi-diurnal tides

9. A type of pyroclastic flow that forms a glassy, melted ash is called
   (a) dolomite
   (b) kryptonite
   (c) ignimbrite
   (d) lucite

10. The difference between a 2-dimensional and 3-dimensional map is that the 3-dimensional also includes
   (a) altitude
   (b) vegetation
   (c) salt wedging
   (d) lava flows

11. Approximately how many thermal springs are there in the United
    States?
    (a) 250
    (b) 500
    (c) 750
    (d) 1000

12. Which of the following is a layer of the atmosphere?
    (a) angiosphere
    (b) thermosphere
    (c) lithosphere
    (d) cyclosphere

13. Fair weather cumulus look like
    (a) Elvis
    (b) horses' tails
    (c) floating cotton balls
    (d) hanging fruit

14. The speed of sound travels through water at around
    (a) 500 m per second
    (b) 1000 m per second
    (c) 1500 m per second
    (d) 2500 m per second

15. *Pillow lava* is formed
    (a) from slow gurgling underwater eruptions
    (b) at one end of bed lava
    (c) only in Antarctica
    (d) of tiny bits of pyroclastic material

16. When a vent opens as a long slit in the ground, it is called a
    (a) central vent
    (b) nuisance
    (c) fissure
    (d) seizure

17. Atmospheric *ozone* is found in a stratospheric layer about
    (a) 2–7 km above the Earth's surface
    (b) 8–12 km above the Earth's surface
    (c) 15–30 km above the Earth's surface
    (d) 30–60 km above the Earth's surface

18. A fault's *dip* or *dip angle* is given by the two measurements
    (a) height and weight
    (b) angle and latitude
    (c) velocity and direction
    (d) angle and direction

19. When many flows of lava pile up on top of each other, forming thick lava plateaus, they are called
    (a) flood basalts
    (b) mountain lava
    (c) fire fountains
    (d) fumaroles

20. The hypocenter of an earthquake is the point at which the
    (a) bells start to ring
    (b) fault strike is formed
    (c) low lying land is flooded
    (d) earthquake slip or rupture starts

21. The central point around which the rest of the hurricane rotates, is called the
    (a) ear
    (b) throat
    (c) troposphere
    (d) eye

22. Frost wedging is a form of
    (a) chemical weathering
    (b) cosmic weathering
    (c) biological weathering
    (d) mechanical weathering

23. Who wrote *The Living Sea*?
    (a) Leonard Nimoy
    (b) Marc Williams
    (c) Jacques Cousteau
    (d) T.R. Gregg

24. What does *Showa Shinsan* mean in Japanese?
    (a) the movie wasn't worth the ticket price
    (b) eat your stir-fry
    (c) earthquake
    (d) new volcano

25. Rock slides take place
    (a) along a clearly visible surface or plane
    (b) only on the ocean floor
    (c) across sandy beaches
    (d) at the same time as snowstorms

26. Seismic waves are primarily caused by
    (a) elephants on a rampage
    (b) vibrations in the earth caused by cracks or shifts in the underlying rock
    (c) cannons firing at football games
    (d) earthworms

27. Of the three ocean density layers, the middle layer is known as the
    (a) deep layer
    (b) thermocline
    (c) pycnocline
    (d) halocline

28. What is a hurricane called in the South Pacific Ocean?
    (a) tornado
    (b) cyclone
    (c) blizzard
    (d) tsunami

29. The recorded vertical distance (altitude) between two neighboring contour lines is called the
    (a) gap
    (b) contour interval
    (c) subduction line
    (d) master joint

30. Mount Erebus, Erta Ale, and Nyiragongo all have
    (a) pillow lava
    (b) extinct volcanoes
    (c) lava lakes
    (d) hard to spell names

31. The map lines that run north and south between the poles are called
    (a) latitude lines
    (b) attitude lines
    (c) longitude lines
    (d) longevity lines

32. What is unusual about the Somali current off the coast of Africa?
    (a) it is cold (4°C) all through the year
    (b) it reverses direction twice a year
    (c) it has an usually large tuna population September through November
    (d) it travels north from South Africa to Morocco all year long

33. What measures the rate of heat loss from exposed skin to that of surrounding air temperatures?
    (a) a thermometer
    (b) wind shear
    (c) wind chill factor
    (d) your mother

34. A very slow (less than 1 cm/year) downhill movement of soil and regolith is known as
    (a) mudflow
    (b) slippage
    (c) soil creep
    (d) dust bowl effect

35. This type of lava moves about 10X slower than basalt and tends to pile up in thick globular deposits
    (a) Aa lava
    (b) rhyolitic lava
    (c) Uh oh lava
    (d) basaltic

36. The Tonga Trench drops the ocean depth nearly
    (a) 1000 m in depth
    (b) 5000 m in depth
    (c) 11,000 m in depth
    (d) 14,000 m in depth

37. What does 1:1,000,000 meters mean when written on a topographical map?
    (a) it is the map's scale
    (b) it is the map's cost
    (c) for every 1 cartographer there are 1 million rocks
    (d) there are 1 million pieces of coal for every 1 diamond

38. This physical weathering type is an important rock breaking force in deserts
    (a) laterite
    (b) biological decay

(c) salt wedging
(d) typhoons

39. Vulcan was the Roman god thought to be the
(a) choreographer to the gods
(b) wine taster to the gods
(c) hair dresser to the gods
(d) blacksmith to the gods

40. On a topographical map, the difference between highs and lows of a land's surface is called the
(a) warf
(b) lineage
(c) relief
(d) focus

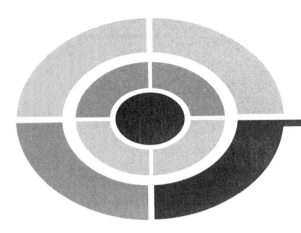

Final Exam

1. Frost wedging is caused by
   (a) the repeated freeze–thaw cycle of water in extreme climates
   (b) pine trees
   (c) ancient impact craters
   (d) cumulus clouds

2. When magma rises to the surface and flows out onto land, it is called
   (a) volcanic steam
   (b) a hot mess
   (c) lava
   (d) glaciation

3. Invertebrates developed and grew the most during this era
   (a) Mesozoic
   (b) Paleozoic
   (c) Cenozoic
   (d) Endozoic

4. The crust is
   (a) something on top of pies
   (b) the thinnest of the Earth's layers
   (c) found around the edge of volcanoes
   (d) the thickest of the Earth's layers

5. How many feet into the crust must be drilled before a 1° Fahrenheit increase in temperature is seen?
   (a) 60
   (b) 75
   (c) 90
   (d) 125

6. The rate of temperature increase compared to depth is known as
   (a) an isothermal gradient
   (b) hypobaric differential
   (c) hot foot gradient
   (d) geothermal gradient

7. When Mount Vesuvius erupted, it covered Pompeii and Herculaneum with chunks of hardened lava and thousands of pounds of
   (a) calcite
   (b) romanite
   (c) malachite
   (d) ignimbrite

8. This era can best be remembered as the era of the dinosaurs
   (a) Paleozoic
   (b) Mesozoic
   (c) Ziozoic
   (d) Cenozoic

9. In general, marble, quartzite, and hornfels have no
   (a) lines of cleavage
   (b) color
   (c) strength
   (d) use

10. Quartzite is the product of metamorphosed sandstone containing mostly
    (a) uranium
    (b) boron
    (c) quartz
    (d) iron

11. Which scientist was the first to suggest that hot magma oozed up through cracks in the sedimentary rock?
    (a) James Hutton
    (b) Evan Hunt
    (c) Charles Darwin
    (d) Alexander Fleming

12. The lines that run horizontally around the globe are called
    (a) attitude lines
    (b) latitude lines
    (c) longitude lines
    (d) longevity lines

13. The upper mantle is also called the
    (a) rind
    (b) asthenosphere
    (c) ionosphere
    (d) coresphere

14. Contours have a set distance used between
    (a) mountain ranges
    (b) axles
    (c) cities
    (d) every contour interval

15. Marble is a metamorphosed limestone or dolomite that has recrystallized after contact with
    (a) low heat
    (b) mercury
    (c) high heat
    (d) meteorites

16. In 1935, a magnitude scale to measure earthquake strength was developed by
    (a) B. McKenna Hunt
    (b) C. S. Lewis
    (c) Charles F. Richter
    (d) Stephen F. Smolen

17. The four major categories, *Protista*, *Fungi*, *Plantae*, and *Animalia* make up what main ICZN group?
    (a) Prokaryotes
    (b) Monera
    (c) Protista
    (d) Eukaryotes

18. A mineral that blocks light completely is called
    (a) transparent
    (b) translucent
    (c) opaque
    (d) a rock

19. What percentage reduced is the *South Atlantic Anomaly's* magnetic field than the majority of the Earth's magnetic field?
    (a) 20%
    (b) 30%
    (c) 40%
    (d) 50%

20. Mineral samples are
    (a) always the same color
    (b) a variety of colors
    (c) always of vitreous luster
    (d) rarely higher than $1\frac{1}{2}$ on the Mohs' scale

21. The radioactive properties of different elements was discovered in 1896, by
    (a) Thomas Edison
    (b) Antoine Becquerel
    (c) Henry Ford
    (d) Paul Edwards

22. Who thought Mother Earth and Father Sky provided for all their needs?
    (a) Native Americans
    (b) Incas
    (c) Neanderthals
    (d) Hawaiians

23. A seamount is
    (a) an extinct ocean volcano
    (b) a tall coral reef
    (c) another name for a seahorse
    (d) a cliff overlooking the ocean

24. Cartography is
    (a) the study of supermarket carts
    (b) the art and science of expressing the physical features of the Earth's surface
    (c) a subset of wagonography
    (d) an unchanging profession

25. When magma forces its way upward into rock, it is known as
    (a) expulsion
    (b) conversion
    (c) transfusion
    (d) intrusion

26. According to the International Code of Zoological Nomenclature, humans are categorized in the superkingdom
    (a) eukaryota
    (b) animalia
    (c) chordata
    (d) sapiens

27. The layer of rock that drifts slowly over the supporting, malleable, upper-mantle layer is called
    (a) magmic rock
    (b) a guyout
    (c) a geological plate
    (d) bedrock

28. Uneven, conchoidal, jagged, and splintery are all types of
    (a) fractures
    (b) lusters
    (c) hardnesses
    (d) streaks

29. The Aurora Borealis and Aurora Australis are found in which atmospheric layer?
    (a) troposphere
    (b) stratosphere
    (c) mesosphere
    (d) thermosphere

30. The place that holds the record for the highest tides in the world is the
    (a) Bay of Pigs
    (b) Mount St. Michel
    (c) San Francisco Bay
    (d) Bay of Fundy

31. Which theory is roughly the reverse of the Contraction theory?
    (a) Convection theory
    (b) Expansion theory
    (c) Contradiction theory
    (d) Subduction theory

32. A contour interval records the
    (a) depth of a lake
    (b) freezing and thawing of rock
    (c) vertical distance (altitude) between two neighboring contour lines
    (d) shape of an Indy 500 racetrack

33. The Animalia category is further divided into
    (a) Vertebrata and Invertebrata
    (b) Vertebrata and Protista
    (c) Invertebrata and Monera
    (d) classical and rock-and-roll

34. Flood, pillow, pahoehoe, and Aa lava are all types of what kind of lava?
    (a) Mediterranean
    (b) Egyptian
    (c) squishy
    (d) basaltic

35. When soil is very porous, with grains not touching each other, the soil is called
    (a) consolidated
    (b) peaty
    (c) unconsolidated
    (d) granite

36. Soil creep takes place as
    (a) the spring rains come
    (b) more and more dirt collects on sweaty skin when hiking
    (c) soil rushes downhill during a rock slide
    (d) a very slow (less than 1 cm/year) downhill movement of soil and regolith

37. Fossils may
    (a) include the location where they were discovered
    (b) be found in the center of the Earth
    (c) be a figment of the movie industry's imagination
    (d) include the outer cast, imprint, or actual remains of a plant or animal preserved in rock

38. The word geode comes from the Greek word *geoides*, which means
    (a) round
    (b) beautiful
    (c) earthlike
    (d) trapped air

39. The atomic mass of magnesium is
    (a) 14
    (b) 22.61
    (c) 24.31
    (d) 86.2

40. Over millions of years, diagenesis transforms sedimentary rock by low-grade
    (a) core transference
    (b) burial metamorphism
    (c) weathering
    (d) glacial movement

41. Which of the following is a layer of the atmosphere?
    (a) blue sphere
    (b) cryosphere
    (c) lithosphere
    (d) stratosphere

42. Vitreous and silky are two different types of
    (a) absolute hardness
    (b) luster
    (c) nodules
    (d) habit

43. How many oxygen molecules does ozone have?
    (a) 1
    (b) 2
    (c) 3
    (d) 4

44. If you can see writing through a mineral, it is
    (a) translucent
    (b) dull
    (c) transparent
    (d) waxy

45. The middle layer separating the lower stratosphere from the thermosphere is called the
    (a) troposphere
    (b) ionosphere
    (c) mesosphere
    (d) outer space

46. Ozone has a distinctive odor and is what color?
    (a) yellow
    (b) tan
    (c) blue
    (d) purple

47. Cinder cones are also called
    (a) vanilla cones
    (b) pine cones
    (c) magma cones
    (d) scoria cones

48. Water
    (a) is a lot heavier than air
    (b) is a lot lighter than air
    (c) about the same as air
    (d) neither heavier nor lighter than air

49. Broken and metamorphic rock fragments found along a meta-morphic rock fault are called
    (a) silt
    (b) deltas
    (c) cinders
    (d) fault breccia

50. What are hurricanes called in the Northwestern Pacific Ocean and Philippines?
    (a) tsunamis
    (b) blue northern
    (c) really big blow
    (d) typhoons

51. Semidiurnal tides take place
    (a) once a day
    (b) twice a day
    (c) once a week
    (d) once every two weeks

52. Lamellar, massive, reniform, and rosette are different types of mineral
    (a) cleavage
    (b) habit
    (c) fracture
    (d) luster

53. *Homo sapiens* are the genus and species name for
    (a) humans
    (b) ant eaters
    (c) mighty lizards
    (d) puppies

54. The only reliable predictor of a volcanic eruption in the hours before an eruption is
    (a) a stillness in the air
    (b) the response of the animal kingdom
    (c) the ringing of the village bell
    (d) a thin trail of smoke from the central vent

55. Which is the largest minerals group?
    (a) carbonates
    (b) silicates
    (c) oxides
    (d) halides

56. The Hawaiian Islands were formed
    (a) from the impact between the Pacific Ocean and North American plates
    (b) at the beginning of time
    (c) over a hot spot
    (d) from a developer's blueprint plans

57. Rainfall and streams in the eastern Rocky Mountains drain to the Atlantic Ocean, while flowing water from the western slopes of the Rocky Mountains runs to the
    (a) Gulf of Mexico
    (b) North Sea
    (c) Pacific Ocean
    (d) Indian Ocean

58. Biomineralization takes place when
    (a) the Moon is closest to the Earth
    (b) calcium carbonate is used to build the shells of sea creatures
    (c) atmospheric carbon moves into the thermosphere
    (d) a dentist fills a cavity

59. Gold fever happens when
    (a) miners get obsessed with the search for gold
    (b) people wear gaudy jewelry
    (c) tooth fillings come loose
    (d) people vacation along the Gold Coast

60. Water covers roughly what % of the Earth's surface?
    (a) 25%
    (b) 45%
    (c) 60%
    (d) 70%

61. Break up of rock happens in all the following ways, except
    (a) frost
    (b) unloading
    (c) lithification
    (d) scarp

62. Temperature ranges in the thermosphere are affected by
    (a) cloud cover
    (b) surface radiation
    (c) ocean temperature
    (d) sun spots and solar flares

63. What kind of eruptions often produce lava lakes?
    (a) Strombolian
    (b) Michaelian
    (c) Hawaiian
    (d) Elisabethan

64. Talc, gypsum, calcite, fluorite, apatite, orthoclase, quartz, topaz, *corundum*, and diamond are standards on the
    (a) absolute scale
    (b) Douglas scale
    (c) Gordon scale
    (d) Mohs' scale

65. Which atmospheric layer can reach temperatures of nearly 2000°C?
    (a) lithosphere
    (b) mesosphere
    (c) thermosphere
    (d) troposphere

66. ICZN stands for
    (a) Ice Cream with Zero Nuts
    (b) International Code of Zookeeping Nomenclature
    (c) Instant Calming of Zany Nerds
    (d) International Code of Zoological Nomenclature

67. The growth of salt crystals, an important rock-breaking force in the desert, is known as
    (a) rock music
    (b) deforestation
    (c) salt wedging
    (d) salinity

68. A person who studies the weather and its atmospheric patterns is called a
    (a) psychologist
    (b) meteorologist
    (c) paleontologist
    (d) cloudologist

69. Earthquakes can signal
    (a) the date swallows return to Capistrano
    (b) movement of magma into vents and channels leading to the surface
    (c) thermal currents in the ocean
    (d) that it's time to think about going on a diet

70. The atmospheric layer closest to the ground is called the
    (a) troposphere
    (b) stratosphere
    (c) mesosphere
    (d) thermosphere

71. When there is a release of internal rock pressure from erosion and the outer layers of a rock are shed, it is known as
    (a) glaciation
    (b) preloading
    (c) hot spots
    (d) unloading

72. One type of cleavage is
    (a) dull
    (b) distinct
    (c) vitreous
    (d) translucent

73. After lava cools and hardens, it becomes
    (a) rock
    (b) red
    (c) edible
    (d) art

74. Dynamothermal metamorphism happens when
    (a) toddlers play together
    (b) temperature and pressure decreases are involved in regional metamorphism
    (c) pressure and temperature increases are involved in regional metamorphism
    (d) sediments collect on the ocean floor

75. The Fujita Wind Damage Scale measures the damage of
    (a) deserts
    (b) hurricanes
    (c) glaciers
    (d) tornadoes

76. Explosions, fiery ash, toxic gases, lava flows, and lahars can all happen when
    (a) you forget to take out the trash
    (b) a volcano erupts
    (c) a lava lake forms a silvery crust
    (d) your boss misses a deadline

77. The ICZN Monera category contains
    (a) elephants and moles
    (b) ferns and pine trees
    (c) bacteria and blue-green algae
    (d) *Allosaurus* and *Apatosaurus*

78. Lenticular clouds are shaped like a
    (a) cow
    (b) ship
    (c) wave
    (d) lens

79. Vibrations in the Earth caused by cracks or shifts in the underlying rock produce
    (a) cirrus clouds
    (b) seismic waves
    (c) increases in water pH
    (d) fish death

80. Gneiss rocks are of what origin?
    (a) sedimentary
    (b) igneous
    (c) metamorphic
    (d) cosmic

81. This type of lava is the most felsic of the different types
    (a) basaltic
    (b) rhyolitic
    (c) dolomitic
    (d) andesitic

82. The location beneath the Earth's surface where a fault rupture begins is called the
    (a) thermocline
    (b) hypocenter
    (c) eye
    (d) epicenter

83. Andrew, Betsy, Camille, Carmen, Gilbert, and Hugo are names of severe
    (a) haircuts
    (b) principals
    (c) hurricanes
    (d) glaciers

84. The study of a rock's physical characteristics through a low-power microscope or hand-held magnifying glass is called
    (a) lithology
    (b) toxicology
    (c) zoology
    (d) virology

85. A key bed is
    (a) difficult to get comfortable on
    (b) a thin, broad, sedimentary bed with specific, easily recognized characteristics
    (c) found only in the Andes Mountains
    (d) always white in color

86. When root growth and burrowing in rock gaps can cause rock fractures to open wider, it is a type of
    (a) temperature weathering
    (b) chemical weathering
    (c) nuclear weathering
    (d) biological weathering

87. The fastest seismic waves and the first to arrive at any monitoring station are the
    (a) s waves
    (b) t waves

(c) o waves

(d) p waves

88. When older fossils are found in the lowest levels of sedimentary rock layers and younger fossils are found closer to the surface, it is called
    (a) mineralization
    (b) fossil succession
    (c) the International Code of Zoological Nomenclature
    (d) granular disintegration

89. What technology do oceanographers use to study the deep ocean floor?
    (a) dredging
    (b) sonar
    (c) fishing
    (d) nanotechnology

90. On April 8, 1906, the San Francisco earthquake struck causing over 1000 deaths, and measuring what magnitude on the Richter scale?
    (a) 3.2
    (b) 5.7
    (c) 8.3
    (d) 9.1

91. The sloshing of water out of a swimming pool, or any body of water, after an earthquake is called
    (a) seiche
    (b) sedimentation
    (c) mud flow
    (d) sink hole

92. The bright red mineral that Renaissance painters used to make red paint pigment is called
    (a) bauxite
    (b) cinnabar
    (c) tourmaline
    (d) jasper

93. What takes place when surface layers are slowly removed from underlying rock?
    (a) mud slide
    (b) volcanic eruption
    (c) denudation
    (d) forest fire

94. Weathering is the
    (a) daily report given by the media to warn of bad weather
    (b) breakdown of rock into smaller pieces of sediment and dissolved minerals
    (c) build up of mid-Atlantic ridge rock
    (d) hardening of organic sediments into rock

95. Seismic moment is figured out through the physics of
    (a) torque
    (b) time
    (c) tongue
    (d) temperature

96. In the SOFAR channel,
    (a) ocean living and gardening is often shown
    (b) low-frequency sounds are not transmitted
    (c) no marine life can exist
    (d) low-frequency sounds can travel for hundreds of kilometers

97. Jacques Cousteau was best known for his work as
    (a) a police inspector, along with his valet, Kato
    (b) an oceanographer and champion of the oceans
    (c) a mime
    (d) a poet

98. Weathering, when combined with gravity, results in
    (a) mud gathering
    (b) heavy top soil
    (c) mass wasting
    (d) clogged waterways

99. Every year southern California experiences about
    (a) 500 earthquakes
    (b) 1000 earthquakes
    (c) 5000 earthquakes
    (d) 10,000 earthquakes

100. What serves as the energy and heat source for the ocean's food chain?
    (a) kelp
    (b) mussels
    (c) sunlight
    (d) sulfur

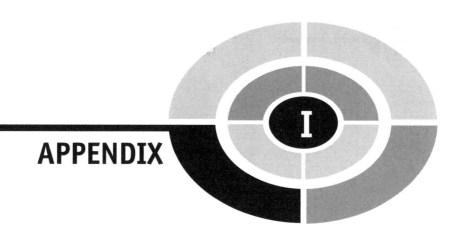

Conversion Factors

LENGTH

1 centimeter	0.3937 inch
1 inch	2.5400 centimeters
1 meter	3.2808 feet; 1.0936 yards
1 foot	0.3048 meter
1 yard	0.9144 meter
1 kilometer	0.6214 (statute); 3281 feet
1 mile (statute)	1.6093 kilometers
1 mile (nautical)	1.8531 kilometers
1 fathom	6 feet; 1.8288 meters
1 angstrom	10^{-8} centimeter
1 micrometer	0.0001 centimeter

VELOCITY

1 kilometer/hour	27.78 centimeters/second
1 mile/hour	17.60 inches/second

AREA

1 square centimeter	0.1550 square inch
1 square inch	6.452 square centimeters
1 square meter	10.764 square feet; 1.1960 square yards
1 square foot	0.0929 square meter
1 square kilometer	0.3861 square mile
1 square mile	2.590 square kilometers
1 acre (USA)	4840 square yards

VOLUME

1 cubic centimeter	0.0610 cubic inch
1 cubic inch	16.3872 cubic centimeters
1 cubic meter	35.314 cubic feet
1 cubic foot	0.02832 cubic meter
1 cubic meter	1.3079 cubic yards
1 cubic yard	0.7646 cubic meter
1 liter	1000 cubic centimeters
1 gallon (USA)	3.7853 liters

MASS

1 gram	0.03527 ounce
1 ounce	28.3496 grams
1 kilogram	2.20462 pounds
1 pound	0.45359 kilogram

PRESSURE

1 kilogram/square centimeter	0.96784 atmosphere; 14.2233 pounds per square inch; 0.98067 bar
1 bar	0.98692 atmosphere; 10^5 pascals

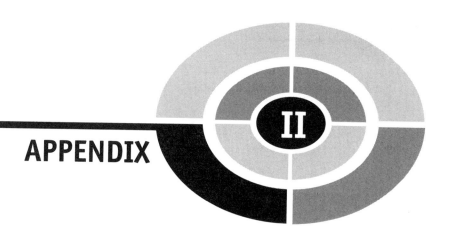

Crystals

Crystal	Crystalline structure	Mohs' hardness	Use *industrial (I)* *ornamental (O)* *gemstone (G)*	Specific gravity
Agate	Hexagonal	6.5–7	O, G	2.6–2.65
Alexandrite	Orthorhombic	8.5	G	3.7
Alamandine	Cubic	6.5–7.5	I, G	4–4.2
Amazonite	Triclinic	6–6.5	O, G	2.56–2.58
Amethyst	Hexagonal	7	O, G	2.63–2.65
Andalusite	Orthorhombic	7–7.5	I, G	3.2
Anglesite	Orthorhombic	2.5–3	O	6.3–6.4

(continued)

Crystal	Crystalline structure	Mohs' hardness	Use *industrial (I) ornamental (O) gemstone (G)*	Specific gravity
Anhydrite	Orthorhombic	3–3.5	I	2.98
Antimony	Trigonal/hexagonal	3–3.5	I	6.7
Apatite	Hexagonal	5	I	3.1–3.2
Aquamarine	Hexagonal	7.5–8	G	2.67–2.7
Argentite	Cubic	2–2.5	I	7.2–7.4
Autunite	Tetragonal	2–2.5	I	3–3.2
Axinite	Triclinic	6.5–7	G	3.4–3.65
Azurite	Monoclinic	3.5–4	I	3.7–3.9
Barite	Orthorhombic	3–3.5	I	4.5
Bauxite	Aggregate of various crystal types	1–3	I	2.0–2.7
Benitoite	Hexagonal	6–6.5	G	3.65–3.68
Bentonite	Monoclinic	1	I	1.2–2.7
Beryl	Hexagonal	7.5–8	G	2.65–2.75
Bismuth	Hexagonal	2–2.5	I	9.7–9.8
Borax	Monoclinic	2–2.5	I	1.7
Brazilianite	Monoclinic	5.5	G	2.9
Brucite	Hexagonal	2.5	I	2.3–2.5
Calcite	Hexagonal	3	I, O, G	2.71
Cassiterite	Tetragonal	6–7	I	6.8–7.1

(*continued*)

Crystal	Crystalline structure	Mohs' hardness	Use *industrial (I)* *ornamental (O)* *gemstone (G)*	Specific gravity
Celestine	Orthorhombic	3–3.5	I, O	3.96–3.98
Cerussite	Orthorhombic	3–3.5	I, G	6.4–6.6
Chalcedony	Hexagonal	6.5–7	G	2.59–2.64
Chalcocite	Orthorhombic	2.5–3	I	5.5–5.8
Chalcopyrite	Tetragonal	3.5–4	I	4.1–4.3
Chromite	Cubic	5.5	I	4.5–4.8
Cinnabar	Hexagonal	2–2.5	I	8.0–8.2
Citrine	Hexagonal	7	G	2.65
Colemanite	Monoclinic	4.5	I, O	2.42
Copper	Cubic	2.5–3	I, O	8.9
Crocoite	Monoclinic	2.5–3	I	6.0
Cuprite	Cubic	3.5–4	I, O, G	6.14
Danburite	Orthorhombic	7–7.5	G	2.97–3
Diamond	Cubic	10	I, G	3.52
Diopside	Monoclinic	5–6	G	3.27–3.31
Dioptase	Hexagonal	5	G	3.3–3.35
Dolomite	Hexagonal	3.5–4	I, O	2.85–2.95
Emerald	Hexagonal	7.5–8	G	2.7–2.78
Enstatite	Orthorhombic	5.5	I	3.26–3.28
Epidote	Monoclinic	6–7	O, G	3.3–3.5

(*continued*)

Crystal	Crystalline structure	Mohs' hardness	Use industrial (I) ornamental (O) gemstone (G)	Specific gravity
Fluorite	Cubic	4	I	3.1–3.3
Galena	Cubic	2.5–3	I, O	7.58
Glauberite	Monoclinic	2.5–3	O	2.8
Goethite	Orthorhombic	5–5.5	I, O	3.3–4.3
Gold	Cubic	2.5	I, O	19.3
Graphite	Hexagonal	1–2	I	2.1–2.3
Grossular	Cubic	6.5–7.5	G	3.6–3.7
Gypsum	Monoclinic	2	I	2.35
Halite	Cubic	2	I	2.1–2.2
Hematite	Hexagonal	5.5–6.5	I, G	5–5.3
Hemimorphite	Orthorhombic	4.5–5	O, G	3.4–3.5
Hiddenite	Monoclinic	6–7	G	3.16–3.2
Indicolite	Hexagonal	7–7.5	G	3–3.25
Iolite	Orthorhombic	7–7.5	G	2.55–2.65
Jadeite (jade)	Monoclinic	6.5–7	O, G	3.33
Jasper	Hexagonal	6.5–7	O	2.6–2.9
Kunzite	Monoclinic	6–7	I, G	3.16–3.2
Kyanite	Triclinic	4–7	I	3.65–3.69
Labradorite	Triclinic	6–6.5	O, G	2.7
Lapis lazuli	Cubic	5–6	O	2.4–2.9

(continued)

Crystal	Crystalline structure	Mohs' hardness	Use industrial (I) ornamental (O) gemstone (G)	Specific gravity
Lazulite	Monoclinic	5–6	G	1.6–1.65
Magnetite	Cubic	5.5–6.5	I	5.2
Malachite	Monoclinic	3.5–4	O	4
Manganite	Monoclinic	4	I, O	4.3
Marcasite	Orthorhombic	6–6.5	O	4.92
Mercury	Hexagonal (below –39°C)	0 (room temperature)	I	13.6
Microlite	Cubic	5–5.5	I, O	4.3–5.7
Molybdenite	Hexagonal	1–1.5	I	4.6–4.7
Monazite	Monoclinic	5–5.5	I	4.6–5.4
Moonstone	Monoclinic	6–6.5	G	2.62
Morganite	Hexagonal	7.5–8	G	2.65–2.75
Muscovite (mica)	Monoclinic	2–3	I	2.77–2.88
Neptunite	Monoclinic	5–6	O	3.19–3.23
Niccolite	Hexagonal	5–5.5	I	7.5–7.8
Olivine (peridot)	Orthorhombic	6.5–7	G	3.27–4.32
Opal	Noncrystalline silica	5.5–6.5	G	1.98–2.1
Orthoclase	Monoclinic	6	I, G	2.5–2.6

(*continued*)

Crystal	Crystalline structure	Mohs' hardness	Use *industrial (I) ornamental (O) gemstone (G)*	Specific gravity
Peridot (chrysolite)	Orthorhombic	6.5–7	G	3.3–4.2
Phenacite	Hexagonal	7.5–8	G	3
Platinum	Cubic	4–4.5	I, O	21.4
Prehnite	Orthorhombic	6–6.5	I	2.8–2.95
Pyrite	Cubic	6–6.5	I	5.02
Pyrrhotite	Hexagonal	3.5–4.5	I	4.6–4.7
Quartz (clear)	Trigonal	7	O, G	2.65
Rhodochrosite	Hexagonal	3.5–4.5	I, O	3.3–3.7
Rhodolite	Cubic	6.5–7.5	O	3.75–3.9
Rhodonite	Triclinic	5.5–6.5	O	3.57–3.76
Rose quartz	Hexagonal	7	O	2.65
Rubellite	Hexagonal	7–7.5	G	3–3.25
Ruby	Hexagonal	9	I, G	4
Sapphire	Hexagonal	9	I, G	4
Scheelite	Tetragonal	4.5–5	I	6.10
Silver	Cubic	2.5–3	I, O	10.5
Smithsonite	Hexagonal	4–5	I, O	4.3–4.5
Sodalite	Cubic	5.5–6	O, G	2.13–2.3
Spessartite	Cubic	6.5–7.5	G	4.2

(continued)

Crystal	Crystalline structure	Mohs' hardness	Use *industrial (I) ornamental (O) gemstone (G)*	Specific gravity
Sphalerite	Cubic	3.5–4	I	3.9–4.2
Spinel	Cubic	8	G	3.6
Staurolite	Monoclinic	7–7.5	G	3.7–3.8
Stibnite	Orthorhombic	2	I	4.6–4.7
Strontianite	Orthorhombic	3.5	I	3.78
Sulfur	Orthorhombic	1.5–2.5	I	2.0–2.1
Sylvite	Cubic	2	I	1.99
Talc	Monoclinic	1	I, O	2.5–2.8
Tanzanite	Orthorhombic	6.5–7	G	3.1–3.4
Tiger eye	Hexagonal	7	O, G	2.65–2.7
Topaz	Orthorhombic	8	G	3.5–3.6
Torbernite	Tetragonal	2–2.5	I	2.68
Tourmaline	Hexagonal	7–7.5	O, G	3–3.25
Turquoise	Triclinic	5–6	O, G	2.6–2.8
Uraninite	Cubic	5–6	I	7.5–10
Variscite	Orthorhombic	4–5	O	2.4–2.6
Vermiculite	Monoclinic	2–3	I	2.4–2.7
Vesuvianite	Tetragonal	6.5	G	3.27–3.45
Zircon	Tetragonal	6.5–7	I, G	4.7

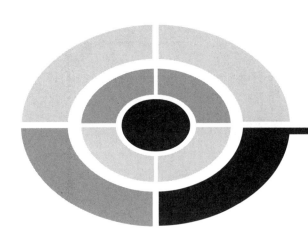

Chapter Quiz Answers

CHAPTER 1

1. C	2. D	3. C	4. B	5. A
6. B	7. B	8. C	9. A	10. D

CHAPTER 2

1. A	2. B	3. B	4. D	5. B
6. D	7. C	8. B	9. A	10. C

CHAPTER 3

1. B	2. A	3. B	4. C	5. D
6. A	7. B	8. D	9. A	10. C

CHAPTER 4

1. B	2. C	3. A	4. B	5. D
6. B	7. D	8. B	9. C	10. A

CHAPTER 5

1. B	2. C	3. A	4. B	5. D
6. D	7. A	8. C	9. C	10. B

CHAPTER 6

| 1. C | 2. C | 3. B | 4. A | 5. C |
| 6. C | 7. D | 8. C | 9. A | 10. B |

CHAPTER 7

| 1. C | 2. D | 3. B | 4. C | 5. C |
| 6. A | 7. B | 8. C | 9. D | 10. A |

CHAPTER 8

| 1. B | 2. D | 3. A | 4. A | 5. C |
| 6. C | 7. B | 8. A | 9. D | 10. B |

CHAPTER 9

| 1. A | 2. B | 3. A | 4. C | 5. B |
| 6. D | 7. D | 8. C | 9. C | 10. B |

CHAPTER 10

| 1. B | 2. C | 3. D | 4. B | 5. B |
| 6. C | 7. A | 8. C | 9. A | 10. D |

CHAPTER 11

| 1. C | 2. B | 3. D | 4. A | 5. B |
| 6. C | 7. C | 8. A | 9. B | 10. D |

CHAPTER 12

| 1. D | 2. A | 3. B | 4. C | 5. B |
| 6. A | 7. D | 8. C | 9. B | 10. C |

CHAPTER 13

| 1. B | 2. C | 3. C | 4. B | 5. D |
| 6. A | 7. B | 8. D | 9. D | 10. C |

CHAPTER 14

| 1. C | 2. B | 3. D | 4. C | 5. C |
| 6. B | 7. D | 8. B | 9. A | 10. B |

CHAPTER 15

| 1. C | 2. B | 3. D | 4. A | 5. A |
| 6. B | 7. D | 8. C | 9. B | 10. A |

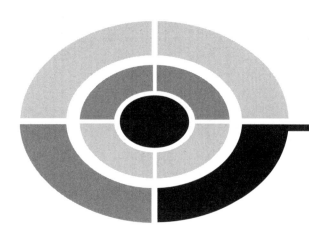

Part Test Answers

PART ONE

1. B	2. C	3. D	4. C	5. D	6. C	7. D	8. A
9. C	10. B	11. A	12. D	13. B	14. D	15. C	16. D
17. B	18. C	19. A	20. C	21. A	22. B	23. B	24. D
25. B	26. D	27. C	28. A	29. B	30. C	31. B	32. D
33. A	34. A	35. A	36. C	37. D	38. A	39. C	40. D

PART TWO

1. B	2. A	3. D	4. A	5. B	6. C	7. B	8. C
9. C	10. D	11. D	12. B	13. A	14. A	15. C	16. A
17. D	18. C	19. B	20. B	21. B	22. C	23. D	24. A
25. B	26. C	27. D	28. B	29. C	30. D	31. B	32. A
33. D	34. B	35. D	36. A	37. C	38. B	39. A	40. C

PART THREE

1. C	2. D	3. B	4. A	5. C	6. B	7. C	8. B
9. C	10. A	11. D	12. B	13. C	14. C	15. A	16. C
17. C	18. D	19. A	20. D	21. D	22. D	23. C	24. D
25. A	26. B	27. C	28. B	29. B	30. C	31. C	32. B
33. C	34. C	35. B	36. C	37. A	38. C	39. D	40. C

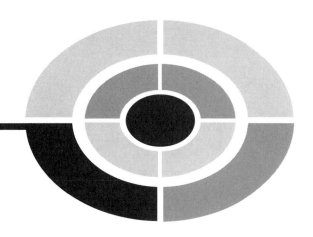

Final Exam Answers

1. A	2. C	3. B	4. B	5. A
6. D	7. D	8. B	9. A	10. C
11. A	12. B	13. B	14. D	15. C
16. C	17. D	18. C	19. B	20. B
21. B	22. A	23. A	24. B	25. D
26. A	27. C	28. A	29. D	30. D
31. B	32. C	33. A	34. D	35. C
36. D	37. D	38. C	39. C	40. B
41. D	42. B	43. C	44. C	45. C
46. C	47. D	48. A	49. D	50. D
51. B	52. B	53. A	54. B	55. B
56. C	57. C	58. B	59. A	60. D
61. C	62. D	63. C	64. D	65. C
66. D	67. C	68. B	69. B	70. A
71. D	72. B	73. A	74. C	75. D
76. B	77. C	78. D	79. B	80. C
81. B	82. B	83. C	84. A	85. B
86. D	87. D	88. B	89. B	90. C
91. A	92. B	93. C	94. B	95. A
96. D	97. B	98. C	99. D	100. C

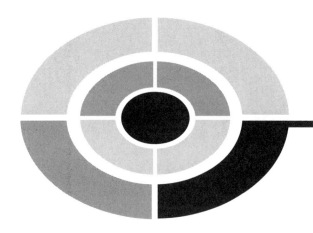

References

Argus, D. and Gordon, R., "Present Tectonic Motion across the Coast Ranges and San Andreas Fault System in Central California," *GSA Bulletin*, 2001, Vol. 113, No. 12, pp. 1580–1592.

Ball, P., *Life's Matrix: A Biography of Water*, Farrar, Straus and Giroux, 2000, New York, NY.

Barrett, P., *Dinosaurs*, National Geographic Society, 2001, Washington, DC.

Benton, M., *Walking with Dinosaurs: Fascinating Facts*, DK Publishing, 2000, New York, NY.

Briggs, D. et al., *The Fossils of the Burgess Shale*, Smithsonian Institute Press, 1994, Washington, DC.

Erickson, J., *Making of the Earth: Geologic Forces that Shape our Planet*, Facts on File, Inc., 2000, New York, NY.

Fisher, R. et al., *Volcanoes: Crucibles of Change*, Princeton University Press, 1997, Princeton, New Jersey.

Fortney, R., *Fossils: The Key to the Past*, Harvard University Press, 1991, Cambridge, Massachusetts.

Gallant, R. A., *Structure: Exploring Earth's Interior*, Benchmark Books, 2003, New York, NY.

Gohau, G., *A History of Geology*, Rutger's University Press, 1990, New Brunswick, New Jersey.

Gordon, R. G., "The Plate Tectonic Approximation: Plate Nonrigidity, Diffuse Plate Boundaries, and Global Plate Reconstructions," *Annu. Rev. Earth Planet. Sci.*, 1998, Vol. 26, pp. 615–642.

Gripp, A. and Gordon, R., "Young Tracks of Hot spots and Current Plate Velocities," *Geophys. J. Ind.*, 2002, Vol. 150, pp. 321–361.

References

Hambrey, M. and Alean, J., *Glaciers*, Press Syndicate of the University of Cambridge, 1992, New York, NY.

Lambert, D., *The Field Guide to Geology*, Facts on File, Inc., Diagram Visual Information, Inc., 1997, New York, NY.

Lambert, D., *The Field Guide to Prehistoric Life*, Facts on File, Inc., Diagram Visual Information, Inc., 1985, New York, NY.

Montgomery, C., *Fundamentals of Geology*, 3rd Ed., William C. Brown Publishers, 1997, Chicago, Illinois.

Mogil, H. M. and Levine, B., *The Amateur Meteorologist: Explorations and Investigations*, Franklin Watts, 1993, New York, NY.

Murphy, B. and Nance, D., *Earth Science Today*, Brooks/Cole Publishing Co., 1999, Pacific Grove, California.

National Academy of Sciences for the American Geological Institute, *Dictionary of Geological Terms*, Dolphin Reference Books Edition, 1962, University of California Press.

Norell, M., *Discovering Dinosaurs in the American Museum of Natural History*, Alfred A. Knopf, 1995, New York, NY.

Pellant, C., *The Complete Book of Rocks and Minerals*, 1995, Dorling Kindersley Publishing, Inc., New York, NY.

Phillips, D. et al., *Blame It on the Weather*, Advantage Publishers Group, 2002, San Diego, CA.

Press, F. and Siever, R., *Understanding Earth*, 2nd Ed., W. H. Freeman and Company, 1998, New York, NY.

Reynolds, Ross, *Guide to the Weather*, Cambridge University Press, 2000, New York, NY.

Skinner, B. and Porter, S., *Physical Geology*, John Wiley and Sons, 1987, New York, NY.

Sofianides, A. and Harlow, G., *Gems and Crystals*, 1990, Simon and Schuster, New York, NY.

Thro, E., *Volcanoes of the United States*, Franklin Watts, 1992, New York, NY.

Van Rose, S., *Earth*, Dorling Kindersley, 2000, New York, NY.

Weidensaul, S., *Fossil Identifier*, Shooting Star Press, 1995, New York, NY.

Williams, J., *The Weather Book*, 1992, Vintage Books, New York, NY.

Wyckoff, J., *Reading the Earth: Landforms in the Making*, Adastra West, Inc., 1999, Mahwah, NJ.

Young, E. and Carruthers, M., *Earth Sciences*, Helicon Publishing, 2001, Oxford, UK.

Internet References

EARTH'S FORMATION

- http://hubble.nasa.gov
- http://science.msfc.nasa.gov
- www.nasa.gov/home/index.html
- http://hubblesite.org/newscenter

EARTHQUAKES

- http://quake.usgs.gov/recenteqs/index.html

FOSSILS

- www.mnh.si.edu
- www.nhm.ac.uk
- www.nationalgeographic.com/features/96/dinoeggs

GLACIERS

- http://www.glacier.rice.edu/
- http://nsidc.org/cryosphere/index.html
- http://www.coolantarctica.com/toc.htm
- http://nsidc.org/data/g01130.html

OCEANS

- http://www.usgs.gov
- http://www.nws.noaa.gov
- www.epa.gov/owow/nps/prevent.html
- http://clean-water.uwex.edu/wav/sds-rcu/sds-rcu.html

SPACE

- http://www.nasa.gov/home/index.html
- http://nssdc.gsfc.nasa.gov/photo_gallery/
- http://earth.jsc.nasa.gov/sseop/efs/

VOLCANOES

- http://vulcan.wr.usgs.gov/Outreach/AboutVolcanoes/how_hot_is_a_ volcano.html

WEATHER/ATMOSPHERE

- www.weather.com
- www.theweathernetwork.com
- http://www.nws.noaa.gov/

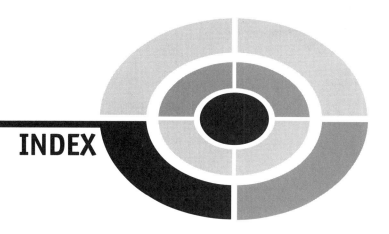

INDEX

Index

Index

Index

Index